THE
LITTLE GREEN
MATH
BOOK

30 POWERFUL PRINCIPLES FOR BUILDING MATH AND NUMERACY SKILLS

Brandon Royal

Published by Maven Publishing

Published by:

Maven Publishing
4520 Manilla Road
Calgary, Alberta, Canada T2G 4B7
www.mavenpublishing.com

Library and Archives Canada Cataloguing in Publication:

Royal, Brandon
The little green math book : 30 powerful principles for building math and numeracy skills / by Brandon Royal.

ISBN 978-1-897393-50-5

Library of Congress Control Number: 2009909355

In addition to the paperback edition, this book is available as an eBook and in the Adobe PDF file format.

Technical Credits:
Cover Design: George Foster, Fairfield, Iowa, USA
Editing: Jonathan K. Cohen, Irvine, California, USA
Illustrations: Ashley Vercekaites, Calgary, Canada

This book's cover text was set in Minion. The interior text was set in Scala and Scala Sans.

Contents

Foreword

Teaching Math in the 1950s

A logger sells a truckload of lumber for $100. His cost of production is four-fifths of his price. What is his profit?

Teaching Math in the 1970s

A logger sells a truckload of lumber for $200. His cost of production is four-fifths of the price, or $160. What is his profit?

Teaching Math in the 1990s

A logger sells a crate of lumber for $400. Her cost of production is $320 and her profit is $80. Your assignment: Please underline the number that represents the profit from the sale.

Teaching Math in the New Millennium

By recycling a crate of pulp and paper, a recycler makes a profit of $100. How do you think the different people involved (for example, loggers, environmentalists, townspeople, wholesalers, consumers) feel about logging and recycling? Class discussion and participation: There are no right or wrong answers!

Quiz

Try these ten basic, but occasionally tricky, math problems. Mark each statement as being either true or false. Answers are found on pages 229–230.

1. If the ratio of females to males at a business conference is 1:2, then the percentage of people who are female at this conference is 50%. ❑ True ❑ False

2. For a given product, markup is always a smaller percentage than margin. ❑ True ❑ False

3. A couple charged $132 on their credit card to pay for a meal while dining out. This $132 figure included a 20% tip which was paid on top of the price of the meal, which already included a sales tax of 10%. The actual price of the meal before tax and tip was $92.40. ❑ True ❑ False

4. Ratios are useful tools that tell us something about actual size or value. ❑ True ❑ False

5. Multiplying a number by 1.2 is the same as dividing that same number by 0.8. ❑ True ❑ False

6. Break-even occurs exactly where profit equals total fixed costs. ❑ True ❑ False

7. A store item that has been discounted first by 20% and then by 30% is now selling at 50% of its original price. ❑ True ❑ False

8. A fast-food vendor registered 100 individual sales during the intermission at a sporting event. The vendor sold 60 hamburgers and 35 hot dogs, and 20 people made purchases which did not include hamburgers or hotdogs. Based on this information, we can conclude that 25 people bought both hamburgers and hot dogs.

    ❑ True   ❑ False

9. If product A is selling for 20% more than product B, then the ratio of product A to product B is 100% to 80%.

    ❑ True   ❑ False

10. Data with a high standard deviation is "bunched." Data with a low standard deviation is more "spread out."

    ❑ True   ❑ False

Introduction

Of the three primary symbols by which we communicate—letters, numbers, and pictures—this book focuses on numbers and, in particular, on "numeracy" skills. Numeracy combines the science of math with the art of numbers. For the everyday student or businessperson, this translates to developing a feel for how numbers work—knowing both the "how" and "why" of problem solving. The ability to perform calculations accurately is one thing; knowing how numbers behave and what makes problems tick is another.

The Little Green Math Book provides a compilation of the most useful academic and real-life math concepts. The secret recipe behind this material lies not only in breaking problems down into topical areas, but in selecting classic problems and presenting relevant tips. Classic problems are recurring, value-added problems which give you, the reader, maximum return for your time and effort spent. Tips serve as principles that underlie math problems; they represent the themes that bind categories of related problems. This book presents 100 classic problems and introduces 30 tips. A three-tier chili rating system helps you gauge a given problem's difficulty level: a one-chili rating indicates that a problem is mild, a two-chili rating signifies that a problem is hot, while a three-chili rating indicates that a problem is very hot.

Good math, like good cuisine, starts with good ingredients.

Chapter 1—Basic Numeracy Skills—highlights the importance of reviewing the basic building blocks of math, which lie at the foundation of superior problem-solving skills. With these skills come the knowledge and confidence to easily answer the following questions: What are the five variations of percentage formulas as they relate to percentage increase and decrease? What pitfalls exist in working with ratios and proportions? Does the order of mathematical operations really matter?

Good math, like good cuisine, often comes with a recipe.

Chapter 2—Wonderful Math Recipes—highlights the importance of knowing the "best" approach. The "group formula" for solving overlap scenarios, the "nine box" table for handling matrix problems, the "barrel method" for solving mixture problems, and the "weighted average" formula for deciphering weighted average problems are all examples of techniques that greatly simplify problem solving.

Good math, like good cuisine, is practical.

Chapter 3—Favorite Numeracy Dishes—highlights math in action. The topic of choice is business and commerce and the topics addressed include the useful topics of markup, margin, cost-price-volume-profit analysis, break-even point, efficiency, and distribution and allocation scenarios.

Good math, like good cuisine, has good presentation.

Chapter 4—Special Math Garnishments—highlights math from a graphical perspective. Charts and graphs are math pictures insofar as they are visual representations of numerical data. Learning how to read graphical information is partly a function of understanding how the type of graph used for presentation purposes—line graph, pie chart, or bar chart—links to the type of data we are trying to summarize. A brief exploration of statistics helps us understand how two variables may be related, how an average is measured in different ways, and what it means if a finding is considered to be statistically significant.

The primary target audience for this book is the everyday student: the high school, community college, and college or university student. Students studying for the math or quantitative sections of College Board tests and standardized entrance exams—including the SAT, ACT, GRE, and GMAT—will find the material especially useful. Given the pervasive nature of math and numeracy skills, this book is also suitable for any person seeking an on-the-job refresher course, particularly professionals pursuing work in banking and management consultancy, accounting, law, and government.

What's special about *The Little Green Math Book?* Teachers and employers frequently complain that students or employees are weak in the basics of math or numeracy skills. Unfortunately, there exists neither a definable, concrete list as to what constitutes the basics nor an efficient method to acquire such skills. This book helps address this deficiency. It provides an answer to two key questions: "What are the math basics?" and "How might a person build basic math and numeracy skills within a reasonable time frame?"

The word "basics" is a relative term. Here, it might be defined as "simple but insightful math" or, more poetically, the point at which a person starts to "dance with math." A rudimentary level of math knowledge is required to solve the majority of problems in this book. Apart from the ability to add, subtract, multiply, and divide, a reader may be called upon to perform basic algebra, including solving for a single variable. That said, even if a person can't recall the formal techniques used to solve basic math problems, the explanations presented for those problems should provide a helpful bridge.

Although the purpose of this material is to help individuals secure a numerical mindset, the greatest benefit to be gained from this book is likely the inspiration to want to continue learning—regardless of the skill area. The satisfaction gained from those "aha" learning moments, coupled with a sense of confidence from taking "ownership" in the material at hand, is a wondrous yet real by-product of the learning process.

Let's get started.

Chapter 1

Basic Numeracy
Ingredients

Math class is tough.
　　　　—Barbie's original
　　　voice chip by Mattel

Percentages

🖝 *Tip #1:*   *A percentage increase is not the same as a percentage of an original number.*

Snapshot

A percentage is a number expressed in relation to 100. The word "percentage" means "hundredths." The percentage sign (%) represents the fraction $\frac{1}{100}$ or its decimal equivalent, 0.01. A percentage is a number that relates two quantities which are usually referred to as the whole and the part.

Below is the most basic percentage formula, as well as two variants:

$$\text{Whole} \times \text{Percent} = \text{Part} \qquad 100 \times 25\% = 25$$

$$\frac{\text{Part}}{\text{Whole}} = \text{Percent} \qquad \frac{25}{100} = 25\%$$

$$\frac{\text{Part}}{\text{Percent}} = \text{Whole} \qquad \frac{25}{25\%} - 100$$

The following problems highlight the five "classic" percentage formulas used to calculate percentage increase and decrease:

(1)   Percentage of an original number

A stock is $100 at the beginning of the year and $120 at the end of the year. The ending value is what percentage of the beginning value?

$$\frac{\text{New}}{\text{Old}} \qquad \frac{\$120}{\$100} = 120\%$$

15

(2)   Percentage increase

A stock is $100 at the beginning of the year and $120 at the end of the year. By what percentage has its value increased?

$$\frac{\text{New} - \text{Old}}{\text{Old}} \qquad \frac{\$120 - \$100}{\$100} = \frac{\$20}{\$100} = 20\%$$

(3)   Percentage decrease

A stock is $100 at the beginning of the year and $80 at the end of the year. By what percentage has its value decreased?

$$\frac{\text{Old} - \text{New}}{\text{Old}} \qquad \frac{\$100 - \$80}{\$100} = \frac{\$20}{\$100} = 20\%$$

(4)   Percentage decrease to return to an original number

A stock is $100 at the beginning of the year and $120 at the end of the year. By what percentage would the stock have to be decreased in order to return to its original price?

$$\frac{\text{New} - \text{Old}}{\text{New}} \qquad \frac{\$120 - \$100}{\$120} = \frac{\$20}{\$120} = 16.67\%$$

(5)   Percentage increase to return to an original number

A stock is $100 at the beginning of the year and $80 at the end of the year. By what percentage would the stock have to be increased in order to return to its original price?

$$\frac{\text{Old} - \text{New}}{\text{New}} \qquad \frac{\$100 - \$80}{\$80} = \frac{\$20}{\$80} = 25\%$$

The following problems test all five variations of percentage formulas.

1. **Antique Stamp ( )**

   The trading value of an antique stamp rose in price from $100 to $150. The current price is what percentage of the original price?

   A)   20%
   B)   25%
   C) $33\frac{1}{3}$%
   D)   50%
   E)   150%

2. **Percentage Increase ( )**

   The trading value of an antique stamp rose in price from $100 to $150. What was the percentage increase in price?

   A)   20%
   B)   25%
   C) $33\frac{1}{3}$%
   D)   50%
   E)   150%

3. **Percentage Decrease ( )**

   The trading value of an antique stamp declined in price from $100 to $50. What was the percentage decrease in price?

   A)   20%
   B)   25%
   C) $33\frac{1}{3}$%
   D)   50%
   E)   150%

4. **Devalue ( )**

The trading value of an antique stamp rose in price from $100 to $125. By what percentage would the price of the stamp have to be decreased in order to return it to its former price?

A)   20%
B)   25%
C) $33\frac{1}{3}\%$
D)   50%
E)   150%

5. **Revalue ( )**

The trading value of an antique stamp declined in price from $100 to $75. By what percentage would the price of the stamp have to be increased in order to return it to its former price?

A)   20%
B)   25%
C) $33\frac{1}{3}\%$
D)   50%
E)   150%

6. Microbrewery (🌶 🌶)

Over the course of a year, a certain microbrewery increased
its output by 70%. At the same time, it decreased its total
working hours by 20%. By what percentage did this factory
increase its output per hour?

A)   50%
B)   90%
C)   112.5%
D)   210%
E)   212.5%

7. Gardener (🌶 🌶)

A gardener increased the length of his rectangle-shaped
garden by increasing its length by 40%, and decreasing its
width by 20%. The area of the new garden

A)   has increased by 20%
B)   has increased by 12%
C)   has increased by 8%
D)   is exactly the same as the old area
E)   cannot be expressed in percentage terms without
     knowing the dimensions of the original garden

☞ *Tip #2:*   *You can't add (or subtract) the percentages of
different wholes.*

Snapshot

There is a fable about a local man who uses a $100 note to place
a bet on his favorite sports team. Unfortunately, on his first bet
he loses $50. He takes his remaining $50 and going for broke, he
wins back $50. Leaving with exactly $100, the same amount he
started with, he says to himself, "Today I'm going home a winner.

I started with $100 and lost 50% of my wealth on the first bet. But betting it all, I increased my wealth by 100% on the second bet. Therefore, I'm overall 50% ahead." Of course, our dear friend is not ahead by 50% because you can't compare (in this case, add) the percentages of different wholes.

Bet #1: $\dfrac{\$100 - \$50}{\$100} = \dfrac{\$50}{\$100} = 50\% \downarrow$

The first bet ends with a 50% decrease in wealth.

Bet #2: $\dfrac{\$100 - \$50}{\$50} = \dfrac{\$50}{\$50} = 100\% \uparrow$

The second bet ends with a 100% increase in wealth.

Compounding of interest presents an everyday example where we cannot simply add percentages together because the base amount (i.e., principal) increases due to compounding. For example, leaving $1,000 in a savings account that pays 10% annual interest would, at the end of two years, yield more than $200 in interest.

Total interest equals:

	Principal × Rate = Interest
First Year:	$1,000 × 10% = $100
Second Year:	$1,100 × 10% = $110
	$210

An alternative, quicker calculation, unfolds as follows:

	Principal × Rate$_1$ × Rate$_2$ = Total
Years 1 & 2:	($1,000 × 110%) × 110% = $1,210
Less Principal:	$1,000
	$210

8. Train ( )

If a passenger train increases its speed by 25% and then increases this speed by 20%, what percentage of the original speed is the increase in speed?

A) 25%
B) 45%
C) 50%
D) 145%
E) 150%

9. Double Discount ( )

A discount of 20% on an order of goods followed by a discount of 10% amounts to

A) the same as one 15% discount
B) the same as one 22% discount
C) the same as one 25% discount
D) the same as one 28% discount
E) the same as one 30% discount

10. Net Effect #1 (✍)

If the price of an item is decreased by 10% and then increased by 10%, the net effect on the price of the item is

A) no change
B) an increase of 1%
C) a decrease of 1%
D) a decrease of 99%
E) variable depending on the actual price involved

11. Net Effect #2 (✍)

If the price of an item is increased by 20% and then decreased by 20%, the net effect on the price of the item is

A) no change
B) an increase of 4%
C) a decrease of 4%
D) a decrease of 96%
E) variable depending on the actual price involved

12. Silver (✍)

The price of silver rose by 100% and then fell by 50%. Compared to the original price, the final price is

A) 50% greater
B) 25% greater
C) the same
D) 25% less
E) 50% less

13. Rule of Finance ( )

A business is growing at 10% a year. How many years will it take for the business to double in size?

A) less than 10 years
B) exactly 10 years
C) greater than 10 years
D) either less than or greater than 10 years
E) it depends on the size of the company

🖝 *Tip #3:* *Percentages cannot be compared directly to numbers because percentages are relative measures while numbers are actual values.*

Snapshot

We generally cannot compare percentages to numbers unless we know the exact numbers represented by the percentages. Percentages, like decimals, fractions, and ratios, are relative measures; actual numbers are absolute values.

Let's presume that in a given company 10% of the employees in department A are salespersons, whereas 20% of the employees in department B are salespersons. Does department B have more salespersons than department A? The answer is that we cannot tell. It could be that both departments have the same number of total employees, in which case department B would definitely have more salespersons than department A (see scenario 1). But there could be many more employees working in department A than in department B. In this case, the number of salespersons working in department A could well be greater than the number of salespersons working in department B (see scenario 2).

Scenario 1

	Dept A	Dept B
Number of Total Employees	100	100
% of Employees Who Are Salespersons	10%	20%
Number of Salespersons	10	20

Scenario 2

	Dept A	Dept B
Number of Total Employees	100	40
% of Employees Who Are Salespersons	10%	20%
Number of Salespersons	10	8

14. Medical School ( )

A student activist group has released a report that suggests that female students have more difficulties in earning admittance to medical school. The facts in the report speak for themselves: 75% of all students in medical school are male, but fewer than half of all female applicants reach their goal of being admitted to medical school.

Which of the following data would be most helpful in evaluating the argument above?

A)   The proportion of all admissions officers who are male versus female.

B)   The percentage of eligible female students admitted to medical school who accept a particular school's admissions offer, and the percentage of eligible male students admitted to medical school who also accept that particular school's admissions offer.

C) The percentage of eligible female students admitted to medical school, and the percentage of eligible male students admitted to medical school.

D) Comparative records on acceptance rates of male and female applicants in other graduate programs such as business, law, and the sciences.

E) The dropout rate for female and male students while in medical school.

15. White-Collar Crime (🐚🐚)

Sociologist Dr. Noah is inconsistent in what he has to tell about the causes of white-collar crime. He claims that unemployment, often aggravated by alcoholism, accounts for at least 50% of the increase in the rate of white-collar crime over the past year. How can he reconcile this with his opinion, expressed earlier in the very same lecture, that the loneliness that results from overwork is at the root of most of the white-collar crime that occurred last year? The author's criticism of Dr. Noah's comments is

A) Warranted, because it cannot be true that most white-collar crime is caused by loneliness if such crime is caused by economic factors.

B) Warranted, because Dr. Noah fails to specify the number of white-collar crimes that occurred ten years ago.

C) Unwarranted, because unemployment and loneliness can both be factors in the same white-collar crime.

D) Unwarranted, because the author fails to mention that severe depression unrelated to unemployment or loneliness can precipitate white-collar crimes.

E) Unwarranted, because there is a difference between the total number of white-collar crimes last year and the number represented by the rate of increase in those crimes last year.

16. Fly Fishing ( )

According to *Anglers Digest* online magazine, the percentage of anglers in Northern Ontario who are using fly-fishing, as opposed to bait fishing, has increased 10% over the last 20 years. If the statements above are true, all of the following statements concerning sports fishing in Northern Ontario during the past two decades could also be true EXCEPT

A) The actual number of people fly-fishing in Northern Ontario has increased while the overall number of people fishing for sport in that region has decreased.

B) The actual number of people fly-fishing in Northern Ontario has increased while the overall number of people fishing for sport in that region has remained the same.

C) The actual number of people fly-fishing in Northern Ontario has increased while the overall number of people fishing for sport in that region has also increased.

D) The actual number of people fly-fishing in Northern Ontario has decreased while the overall number of people fishing for sport in that region has also decreased.

E) The actual number of people fly-fishing in Northern Ontario has remained the same while the overall number of people fishing for sport in that region has increased.

>☞ *Tip #4:* *Percentages (or ratios) are necessary tools for drawing conclusions about qualitative measures, such as preferability, danger, or safety.*

Snapshot

Is train travel becoming more dangerous? Even if, in a given year, there are more train crashes and regrettably more injuries/deaths compared with a previous year, this does not necessarily mean that train travel is becoming more dangerous. We have to examine how many people are riding on trains (or, more precisely, how many injuries/deaths per train mile logged). If many more people are taking trains, it is likely that the percentage or ratio of train related injuries/deaths (to total passengers) is decreasing. In such a case, train travel would be considered safer, not more dangerous. Ratios (or fractions, decimals, or percents) are necessary tools for drawing conclusions about qualitative measures such as preferability, danger, or safety.

17. Lottery (𝄞)

Income	Number of ticket purchases
Less than $20,000	10,000
$20,001 – $50,000	7,000
$50,001 – $100,000	4,000
Over $100,000	800

The data above, based on a recent report prepared by Lotto Corp., indicates that the number of people who spend $1 to buy a lottery ticket is found to be related to their average annual income. Assuming that this survey was representative of the buying patterns of individuals within each income category, are people who make more money less likely to buy lottery tickets?

A)   Yes
B)   No
C)   Can't Tell

18. Iraq (ʃ ʃ)

During the first three years of the Second Gulf War, 2,300 members of the United States Armed Forces died fighting in Iraq. During this same three-year period, some 100,000 civilians died in the United States from automobile-related accidents. These figures reveal a sad fact: it is more dangerous for a civilian to drive on the streets and highways of America than it is to be dressed in uniform and serving in Iraq.

Which of the following calculations, during this same three-year period, would reveal most clearly the absurdity of the conclusion drawn above?

A)   Comparing deaths caused by combat in the U.S. armed forces in Iraq to gun-related deaths in the domestic U.S.

B)   Comparing the number of deaths among U.S. civilians that resulted from driving at high speeds to the total number of deaths among members of the armed forces.

C)   Factoring out deaths caused by accidents during service in the armed forces in Iraq as well as those due to traffic deaths in the U.S. which were not considered to be the fault of those individuals.

D) Comparing death rates per thousand members of each group rather than total number of deaths.

E) Factoring into consideration the small percentage of time that the average driver in the U.S. drives his or her car each day compared with the full-time status of members of the armed forces in Iraq.

☞ *Tip #5:* *Growing pies presents a classic case where the dollar value of a given item increases over time, yet this same item decreases in percentage terms relative to a larger pie.*

Snapshot

Growing or shrinking pies present examples of the tangling of numbers and percents.

Growing Pie—You may spend more money on personal phone calls this year compared with last year, but this does not mean that the percentage of your disposable income that you spend on phone calls has gone up. It is quite possible, for example, that you have enjoyed a significant increase in your disposable income. Therefore, although the actual money that you spend on personal phone calls has gone up, the percentage of your disposable income that you spend on phone calls has gone down. In short, a growing pie often results in an increase in dollar expenditures despite a corresponding decrease in the percentage of money being spent.

Shrinking Pie—You may spend fewer hours dining out than you used to, but this does not mean that you spend less of your free time dining out. What if you have a lot less free time these days? This would imply that an equal (or even smaller) number of hours dining out would mean that a greater portion (percentage) of your free time is spent dining out. In short, a shrinking pie often results in a decrease in dollar expenditures despite a corresponding increase in the percentage of money being spent.

19. Military Expenditures (𝕁 𝕁)

Proponents of greater military spending for the Reno Republic argue that the portion of their national budget devoted to military programs has been declining steadily for a number of years, largely because of rising infrastructure development costs. Yet groups opposed to increasing military expenditure point out that the Reno Republic military budget, even when measured in constant, or inflation-adjusted, dollars, has increased every year for the past two decades. Which of the following, if true, best resolves the apparent contradiction presented above?

A) Countries in the same region as the Reno Republic have increased the annual portion of their national budgets devoted to military spending.

B) The total national budget has increased faster than military expenditure has increased.

C) The advocates of greater military spending have overestimated the amount needed for adequate defense of the country.

D) Military expenditures have risen at a rate higher than the rate of inflation.

E) Total military expenditures have increased faster than the national budget has increased.

☞ *Tip #6:*   *Consider the following statement: "Person A spends a greater portion of his or her disposable income on beer than on coffee as compared to person B." This does not necessarily mean that person A spends more money in total on beer and coffee than does person B. It also does not mean that person A, individually, spends more money on beer than he or she does on coffee.*

Snapshot

To preview the concept tested in the upcoming problem, titled *Fiction Books,* ponder the following statement:

The percentage of white wine to red wine bottled in Germany is greater than the percentage of white wine bottled in France.

Does this mean that there is more white wine bottled in Germany than white wine bottled in France? Answer—No. Why? Because the size of the wine industry in France is so much bigger than that of the wine industry in Germany. The French "pie" is much bigger than the German "pie," five times bigger, in fact.

Does the original statement mean that Germany bottles more white wine than it does red wine or that France bottles more white wine than it does red wine? The answer is no on both accounts.

Let's draw upon some plausible numbers in line with statistics from the wine industry. France produces five times as much wine as Germany does. Of all the wine produced by Germany, 75% is white and 25% is red. Of all the wine produced by France, 25% is white and 75% is red. Even though the percentage of white wine to red wine bottled in Germany is greater than the percentage of white wine bottled in France, France still produces more white wine than Germany does. Here are some simple numbers to prove our point. Say that France produces 500 bottles of wine each year and Germany produces 100 bottles of wine each year. This means that France produces 125 bottles of white wine whereas Germany produces 75 bottles of white wine.

French Wine Production 500 bottles		German Wine Production 100 bottles	
(White Wine) 125 bottles	(Red Wine) 375 bottles	(White Wine) 75 bottles	(Red Wine) 25 bottles

20. Fiction Books (𝄞𝄞𝄞)

A major publishing conglomerate released a survey concerning the relationship between a household's education level and the kind of books found in its library. Specifically, members of households with higher levels of education had more books in their libraries. The survey also indicated that compared with members of households with lower levels of education, members of households with higher levels of education had a greater percentage of fiction versus nonfiction books.

Which of the following can be properly inferred from the survey results cited above?

A) People with the highest levels of education buy more fiction books than nonfiction books.

B) Members of households with higher levels of education have more fiction books than nonfiction books.

C) Members of households with lower levels of education have more nonfiction books than fiction books.

D)   Members of households with lower levels of education have more nonfiction books than do members of households with higher levels of education.

E)   Members of households with higher levels of education have more fiction books than do members of households with lower levels of education.

Ratios and Proportions

☞ *Tip #7:*   *Ratios tell us nothing about actual size or value.*

Snapshot

Ratios

A ratio is a comparison. There are two ways to write a ratio. The first uses the word "to." The second uses a colon. For example:

7 to 8   or   7 : 8

The above representations have identical meanings.

We can and should always reduce a ratio. So 4:8 is written 1:2. Ratios should be expressed using integers, not fractions or decimals.

Example $\frac{1}{2} : 2 \rightarrow 1 : 4$ $1 : 1.5 \rightarrow 2 : 3$

Ratios are very useful for making relative comparisons, especially when two items differ greatly in size.

"Three out of every five Europeans have a passport while only one out of every five Americans have a passport."

Here we don't have to worry about how large the U.S. population is compared with that of Europe because a ratio allows us to make a relative comparison.

Now assume for a moment that you hear of two businesses that both claim to have two dollars in cash for every one dollar owed to creditors. Can we conclude that these businesses have a lot of excess cash? The answer is that we can't tell. This simple example highlights an inherent weakness of ratios—they obscure size because they disguise actual values.

Organization	Ratio of assets to liabilities	Dollar size
Business A	1 to 2	With $2 billion in cash (or monetary equivalents) and $1 billion in liabilities, Super Corp. still has $1 billion in excess cash.
Business B	1 to 2	With $2,000 in cash and $1,000 in liabilities, Keo's Kiosk has $1,000 in excess cash.

Proportions

A proportion is an equation which shows that two ratios are equal. Equivalent or equal fractions form a proportion. There are two ways to write a proportion. The first uses four dots: "::". The second uses an "=" sign to express a proportion as an equation. Thus, the following are equivalent:

Example $\quad \dfrac{1}{2} :: \dfrac{50}{100} \qquad \dfrac{1}{2} = \dfrac{50}{100}$

In a proportion, cross products are equal:

Example $\quad \dfrac{a}{b} :: \dfrac{c}{d} \qquad a \times d = b \times c$

Example $\dfrac{1}{2} :: \dfrac{50}{100}$ $1 \times 100 = 2 \times 50$

Solving proportions as equations:

$$\dfrac{3}{5} = \dfrac{x}{50}$$

$$3 \times 50 = 5x$$

$$150 = 5x$$

$$5x = 150$$

$$x = \dfrac{150}{5}$$

$$x = 30$$

There are four ways to structure proportions:

$\dfrac{\text{Part}}{\text{Whole}} :: \dfrac{\text{Part}}{\text{Whole}}$ Example $\dfrac{3}{10} :: \dfrac{300}{1000}$

or

$\dfrac{\text{Whole}}{\text{Part}} :: \dfrac{\text{Whole}}{\text{Part}}$ Example $\dfrac{10}{3} :: \dfrac{1000}{300}$

or

$\dfrac{\text{Part}}{\text{Part}} :: \dfrac{\text{Whole}}{\text{Whole}}$ Example $\dfrac{3}{300} :: \dfrac{10}{1000}$

or

$\dfrac{\text{Whole}}{\text{Whole}} :: \dfrac{\text{Part}}{\text{Part}}$ Example $\dfrac{10}{1000} :: \dfrac{3}{300}$

21. WAA (𝄞𝄞)

The World Automobile Association (WAA) publishes a list of the "Best and Worst Drivers of the World," ranking the drivers of every nation according to the number of traffic deaths per mile driven in that country.

Each of the following, if true, would by itself provide a logical objection to using the WAA's ranking as a representation of the quality of drivers in each nation EXCEPT

A)   Some countries contain hundreds of thousands of miles of road while other countries contain relatively few miles of road.

B)   The roads in some countries are in bad repair and are therefore more dangerous than roads in other countries.

C)   The average driver in industrialized countries can afford to maintain his or her car in better condition than can the average driver in less developed countries.

D)   Minor accidents that would cause little injury in many countries are often fatal when they occur in extremely mountainous countries.

E)   Because of differences in national economies, the average car in some countries contains many more passengers than does the average car in other countries.

> ☛ *Tip #8:*   *You cannot add or subtract a number from a ratio unless you know the exact numbers that make up the ratio. However, you can always multiply or divide both sides of a ratio by a given number.*

Snapshot

Ponder the following: "The ratio of white goldfish to orange goldfish in a tank is 3:10. If two white goldfish are added to the tank, what is the new ratio of white to orange goldfish?"

If your instinct is to blurt out, "50 percent," you're not alone. After all, if we add 2 and 3 to get 5, we can create a new ratio of 5:10. This simplifies to 1:2 or 50 percent. You might, on the other hand, be second-guessing yourself, wondering whether the answer is 5:12. After all, 2 + 3 = 5 and 2 + 10 = 12. This answer, however, is clearly flawed because we could never just mathematically add two white goldfish to the number of orange gold fish.

The answer is that we can't tell. It could be that the ratio is indeed 5:10 or 50 percent. But this assumes that there are 3 white goldfish and 10 orange goldfish to begin with. What if there are 3,000 white goldfish and 10,000 orange goldfish? Then the ratio of 3 to 10 white to orange goldfish will hardly change when two white goldfish are added to the tank.

The moral here is that we can only add or subtract from a ratio when we know the exact numbers that make up the ratio.

22. Scotch & Soda (🌶)

An alcoholic mixture that is 1 part scotch to 2 parts soda is added to another mixture that is 2 parts scotch to 3 parts soda. What is the ratio of scotch to soda in the resulting mixture?

A)   3:5
B)   3:8
C)   5:3
D)   8:3
E)   Cannot be determined

☞ *Tip #9:*   *A part-to-part ratio is not the same thing as a part-to-whole ratio. If the ratio of female to male attendees at a conference is 1:2, this does not mean that 50 percent of the conference attendees are female.*

Snapshot

If the ratio of women to men at a conference is 1 to 2, what percentage of the conference is female? The answer is not 50%, but instead $33\frac{1}{3}\%$. That is, the ratio of women to total attendees is 1:3 or $33\frac{1}{3}\%$. It is easy to confuse part-to-part ratios with part-to-whole ratios. The ratio of women to men is a part-to-part ratio; the ratio of women to total attendees is a part-to-whole ratio. Just substitute some numbers to make everything clear. Say there are 5 females and 10 males at the conference. What percentage of the conference attendees are female? Answer: $\frac{5}{15} = \frac{1}{3} = 33\frac{1}{3}\%$.

Ratio of women to men at the conference	Ratio of women to total attendees at the conference	Fraction of women to total attendees at the conference	Percentage of women to total attendees at the conference
1 to 2	1 to 3	$\frac{1}{3}$	$33\frac{1}{3}\%$

23. Eggs (🌶)

In a box of one dozen eggs, the ratio of broken eggs to unbroken eggs could be any of the following EXCEPT

A)   1 : 1
B)   1 : 2
C)   1 : 3
D)   1 : 4
E)   1 : 5

24. Elk Herd (🌶)

If the ratio of male elk to female elk in a herd is $16 : 9$, what fraction of the herd is male?

A) $\frac{16}{9}$ B) $\frac{16}{25}$ C) $\frac{9}{16}$ D) $\frac{9}{25}$ E) $\frac{1}{4}$

25. Rum & Coke (🌶️🌶️)

A glass holding 6 ounces of an alcoholic drink that is 1 part rum to 2 parts Coke is added to a jug holding 32 ounces of an alcoholic drink that is 1 part rum to 3 parts Coke. What is the ratio of rum to Coke in the resulting mixture?

A)   2 : 5
B)   5 : 14
C)   3 : 5
D)   4 : 7
E)   14 : 5

26. Toledo Prep (🌶️🌶️)

At Toledo Prep School, the faculty-to-student ratio is 1 : 7. If 3 of every 7 students are female and one-fifth of the faculty is female, what fraction of the combined students and faculty is female?

A)   30%
B)   32%
C)   40%
D)   42%
E)   57%

27. Fuchsia (🌶🌶🌶)

At Deluxe paint store, fuchsia paint is made by mixing 5 parts of red paint with 3 parts of blue paint. Mauve paint is made by mixing 3 parts of red paint with 5 parts blue paint. How many liters of blue paint must be added to 24 liters of fuchsia to change it to mauve paint?

A) 9
B) 12
C) 15
D) 16
E) 18

☞ *Tip #10:* *Proportions, in the form of $\frac{A}{B} = \frac{C}{D}$, are among the simplest but most versatile math problem-solving tools.*

Snapshot

On a certain trading day, one Australian dollar could be exchanged for 0.70 Canadian dollars. How many Australian dollar could be exchanged for 280 Canadian dollars?

Ratio Structure:

$$\frac{\text{Part}}{\text{Whole}} :: \frac{\text{Part}}{\text{Whole}}$$

Applied to this problem:

$$\frac{\text{Australian dollar}}{\text{Canadian dollar}} :: \frac{\text{Australian dollar}}{\text{Canadian dollar}}$$

Calculation:

$$\frac{1_{\text{AUD\$}}}{0.80_{\text{CAN\$}}} = \frac{x_{\text{AUD\$}}}{240_{\text{CAN\$}}}$$

$$1(240) = x(0.80)$$

$$x(0.80) = 1(240)$$

$$\frac{1}{0.80} \times x(0.80) = \frac{1}{0.80} \times 1(240)$$

$$x = \frac{240}{0.80}$$

$x = 300$ Australian dollars

28. Exchange (🌶)

If one Philippine peso can be exchanged for 0.15 Hong Kong dollars, then how many pesos will a person receive from exchanging one Hong Kong dollar?

A) 0.15
B) 0.67
C) 1.50
D) 6.67
E) 15.0

29. Sale Time (🌶)

During a holiday special, a customer buys a pair of shoes for $12 off the regular retail price. If all items were marked 30% off, what was the original price of the shoes?

A) $15.60
B) $28.00
C) $36.00
D) $40.00
E) $52.00

30. Shrinkage ( )

The manager of "I Scream for Summer," an outdoor ice cream kiosk, must make an allowance for an average 25% loss of overall weight that occurs when ice cream melts, evaporates, or spoils in hot weather. If this 25% loss figure is accurate, how many ounces of ice cream (on average) are needed to yield a deluxe 24-ounce dish of ice cream?

A)   18
B)   30
C)   32
D)   36
E)   40

☞ Tip #11: *If product A is selling for 20% more than product B, then the ratio of product A to product B is 120% to 100% (not 100% to 80%). If product B is selling for 20% less than product A, then the ratio of product A to product B is 100% to 80% (not 120% to 100%).*

Snapshot

When visualizing percent problems, relate everything to a 100 percent base. If we have a 20% increase, this means that we started at 100% and have moved to 120% ($\frac{120\% - 100\%}{100\%} = \frac{20\%}{100\%} = 20\%$). Note that it is not the same as starting at 80% and moving to 100%. After all, going from 80% to 100% represents an increase of 25%, not 20% ($\frac{100\% - 80\%}{80\%} = \frac{20\%}{80\%} = 25\%$).

Likewise, if something decreases by 20%, we visualize starting at 100% and moving down to 80% ($\frac{100\% - 80\%}{100\%} = \frac{20\%}{100\%} = 20\%$). This is not the same as moving from 120% to 100%. After all, going from 120% to 100% represents a decrease of $16\frac{2}{3}\%$, not 20% $(\frac{120\% - 100\%}{120\%} = \frac{20\%}{120\%} = 16\frac{2}{3}\%)$.

31. Product A ( )

Product A sells for 20% more than product B, and product B sells for $40. What is the price of product A?

A) $32
B) $40
C) $41.67
D) $48
E) $50

32. Product B ( )

Product A sells for 20% more than product B, and product A sells for $50. What is the price of product B?

A) $32
B) $40
C) $41.67
D) $48
E) $50

33. Product C ( )

Product D sells for 20% less than product C, and product D sells for $40. What is the price of product C?

A) $32
B) $40
C) $41.67
D) $48
E) $50

34. Product D (🌶)

Product D sells for 20% less than product C, and product C sells for $50. What is the price of product D?

A) $32
B) $40
C) $41.67
D) $48
E) $50

35. Diners (🌶🌶)

A man charged $264 on his credit card in payment for a family dinner. This $264 figure included a 20% tip which was paid on top of the price of the food, which already included a sales tax of 10%. What was the actual price of the meal before tax and tip?

A) $185
B) $200
C) $203
D) $206
E) $216

36. Investments (🌶️🌶️)

A lady sold two small investment properties, A and B, for $24,000 each. If she sold property A for 20% more than she paid for it, and sold property B for 20% less than she paid for it, then, in terms of the net financial effect of these two investments (excluding taxes and expenses), we can conclude that the lady

A)   broke even
B)   had an overall gain of $1,200
C)   had an overall loss of $1,200
D)   had an overall gain of $2,000
E)   had an overall loss of $2,000

Reciprocals

☞ *Tip #12:   Dividing by any number produces the same result as multiplying by that number's reciprocal (and vice-versa).*

Snapshot

Informally speaking, the reciprocal of a number is that number "upside down." Technically, the reciprocal (or multiplicative inverse) of a number is another number, which when multiplied by the original number, yields a product of 1.

Ex. $7 \times \dfrac{1}{7} = 1$ $\dfrac{1}{7}$ is the reciprocal of 7 (and vice versa)

Ex. $\dfrac{1}{2} \times \dfrac{2}{1} = 1$ $\dfrac{2}{1}$ is the reciprocal of $\dfrac{1}{2}$ (and vice versa)

Ex. $\dfrac{3}{4} \times \dfrac{4}{3} = 1$ $\dfrac{4}{3}$ is the reciprocal of $\dfrac{3}{4}$ (and vice versa)

Scenario 1: Multiplying by any number is exactly the same as dividing by that number's reciprocal. For example, multiplying by 2 is the same as dividing by $\frac{1}{2}$.

Ex. $4 \times 2 = 8$ or $4 \times 2 = 4 \div \frac{1}{2} = \frac{\frac{4}{1}}{\frac{1}{2}} = 8$

Scenario 2: Dividing by a number is exactly the same as multiplying by that number's reciprocal. For example, dividing by 2 is the same as multiplying by $\frac{1}{2}$.

Ex. $4 \div 2 = 2$ or $4 \div 2 = \frac{4}{1} \times \frac{1}{2} = 2$

37. Deceptive (🌶)

Dividing 100 by 0.75 will lead to the same mathematical result as multiplying 100 by which number?

A)   0.25
B)   0.75
C)   1.25
D)   1.33
E)   1.75

38. Topsy-Turvy (🌶)

Multiplying $100 by 1.2 is equal to $120. Instead of multiplying $100 by 1.2 to get $120, we could have also divided the original $100 by _____ to get $120.

A) 0.20
B) 0.80
C) 0.83
D) 1.16
E) 1.20

Order of Operations

👉 *Tip #13:* *The order in which two or more numbers are added or multiplied does not matter. However, the order in which two or more numbers are subtracted or divided does matter.*

Snapshot

The commutative law and the associative law are two mathematical laws of operation which are easy to confuse.

Commutative Law

The commutative law states that order doesn't matter. This law holds for addition and multiplication, but <u>not</u> for subtraction or division.

Addition:
$$a + b = b + a$$
$$2 + 3 = 3 + 2$$

Multiplication:

$$a \times b = b \times a$$
$$2 \times 3 = 3 \times 2$$

BUT NOT:

Subtraction:

$$a - b \neq b - a$$
$$4 - 2 \neq 2 - 4$$

Division:

$$a \div b \neq b \div a$$
$$4 \div 2 \neq 2 \div 4$$

Associative Law

The associative law states that regrouping doesn't matter. This law holds for addition and multiplication, but <u>not</u> for subtraction or division.

Addition:

$$(a + b) + c = a + (b + c)$$
$$(2 + 3) + 4 = 2 + (3 + 4)$$

Multiplication:

$$(a \times b) \times c = a \times (b \times c)$$
$$(2 \times 3) \times 4 = 2 \times (3 \times 4)$$

BUT NOT:

Subtraction:

$$(a - b) - c \neq a - (b - c)$$
$$(8 - 2) - 4 \neq 8 - (2 - 4)$$

Division:

$$(a \div b) \div c \neq a \div (b \div c)$$
$$(8 \div 2) \div 4 \neq 8 \div (2 \div 4)$$

39. Partners ( )

If a, b, c, and d are real numbers, each of the following expressions equals $a(bcd)$ EXCEPT

A) $(dcb)a$
B) $d(abc)$
C) $(ab)(cd)$
D) $(ad)(bc)$
E) $(ab)(ac)(ad)$

40. Bargain ( )

A discount of 10% on an order of goods followed by a discount of 30% amounts to

A) the same as one 33% discount
B) the same as one 40% discount
C) the same as a discount of 30% followed by a discount of 10%
D) less than a discount of 30% followed by a discount of 10%
E) more than a discount of 30% followed by a discount of 10%

Chapter 2

Wonderful Math Recipes

Four out of every three persons
have trouble with fractions.
                    —Anonymous

Overlap Scenarios

🖝 *Tip #14:* *Situations involving either A or B do not necessarily preclude the possibility of both A and B. When two events overlap, the number of items in both groups will exceed 100% due to mutual inclusivity.*

Snapshot

Group problems (also called overlap problems or Venn-diagram problems) illustrate situations in which there is crossover among sets of information. These types of problems may be thought of as membership problems—some members do this, some members do that, some members do both, and some members do neither.

Consider the following exchange:

Travel Agent: Of people who booked through our agency and traveled to Australia or New Zealand this past winter, 65% went to Australia and 55% went to New Zealand.

Customer: That doesn't sound right. You should check your numbers. How can more than 100% of your tourists travel to Australia or New Zealand?

The customer's response is likely invalid, because it is entirely possible that some tourists traveled to both Australia and New Zealand during their winter vacation.

How can we work with this kind of information in order to get a clear picture of the number of people in each category? The key lies in manipulating the following formula:

Two-Groups Formula		Solve
Add:	Members in Group A	x
Add:	Members in Group B	x
Less:	Both	$<x>$
Add:	Neither	x
	Total	xx

In Las Vegas, 21 tourists were surveyed, and it was found that 15 of these people liked to play the slot machines and 7 people liked to play table games such as blackjack. Only 4 people surveyed said they played neither slot machines nor table games. How many people surveyed played both slot machines and table games?

Two-Groups Formula		Solve
Add:	Slot Machines	15
Add:	Table Games	7
Less:	Both	$<?>$
Add:	Neither	4
	Total	21

Answer: Five people played both slot machines and table games.

Here is the visual representation of the Las Vegas scenario.

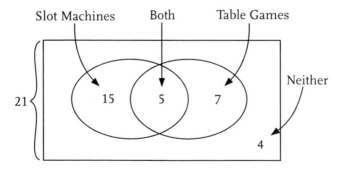

Calculation:

Group A + Group B – Both + Neither = Total
15 + 7 – x + 4 = 21
26 – x = 21
x = 5

The rectangular region represents 100 percent—the whole pie, so to speak. The total of all the numbers in the rectangle cannot add up to more than 21 people. This is why "neither" must be added back; it falls outside the circle but within the rectangle. "Both" must be subtracted out because it was counted twice, once when counting members of Group A and again when counting members of Group B.

Summary of Formula Templates

Two-Groups Formula		Solve	Three-Groups Formula		Solve
Add:	Group A	x	Add:	Group A	x
Add:	Group B	x	Add:	Group B	x
Less:	Both	<x>	Add:	Group C	x
Add:	Neither	x	Less:	A & B	<x>
	Total	xx	Less:	A & C	<x>
			Less:	B & C	<x>
			Add:	All of A & B & C	x
			Add:	None of A or B or C	x
				Total	xx

Author's note:

Let's briefly contrast what is commonly referred to as the "inclusive or" versus the "exclusive or." Consider the statement "Today's weather will be rainy or windy." In casual conversation this may be interpreted to mean that the weather is anticipated to be either rainy or windy, but not both rainy and windy. In logic parlance, however, such situations involving either A or B do not necessarily preclude the possibility of both A and B. In other words, the weather could indeed be both rainy and windy. There is an implied "both" in this particular type of "either-or" statement. The multiple-choice problems encountered in this section also involve the use of the "inclusive or." That is, these problems provide for the possibility of both A and B.

Other situations may involve the use of the "exclusive or." Here it is not possible to have a situation involving both A and B. Take for example the student who must choose to enroll in either high school A or high school B. Obviously the student cannot enroll at both schools. We effectively understand that the choice is between A or B, and does not include both A and B. Likewise, the negative statement "We're not going to visit Italy or Spain during our summer trip" must be logically interpreted as "We're not going to Italy or Spain, and certainly not to both Italy and Spain."

The same exclusivity holds among negative statements. Consider the sign: "No eating or drinking allowed." We cannot interpret this sign to mean that because we can neither eat nor drink, we can therefore eat <u>and</u> drink simultaneously in order to avoid the "no eating <u>or</u> drinking" mandate.

In summary, situations involving the use of the "exclusive or" lead to *mutually exclusive* options. Situations involving the use of the "inclusive or" lead to *mutually inclusive* options. Note also that we may refer to two options that are mutually inclusive as two options that are not mutually exclusive.

41. Country Club (🌶)

A country club has 2,500 members, 700 of whom play tennis and 500 of whom play golf. If 1,500 members play neither tennis nor golf, how many members play both tennis and golf?

A)   200
B)   1,000
C)   1,300
D)   2,200
E)   2,800

42. Nova vs. Rebound (🌶🌶)

Jill: In our survey, 53% of those questioned said they used Nova Running Shoes for jogging and 47% said they used Rebound Running Shoes for jogging.

Jack: In that case everyone questioned used one product or the other.

Jill: No, 24% said they didn't use either one.

Jill's statements imply that

A)   Not all of those questioned liked to go jogging.

B)   Some of those who said they used either Nova or Rebound Running Shoes for jogging were lying.

C)   Some of those who used neither product were nevertheless familiar with Nova and Rebound Running Shoes.

D)   Some of those who used Nova Running Shoes for jogging were among those who also used Rebound Running Shoes for jogging.

E)   Many Nova and Rebound Running Shoes are actually made in the same factories in various international locations.

43. Science (ꝑ)

Each of 35 students is enrolled in chemistry, physics, or both. If 22 are enrolled in chemistry and 19 are enrolled in physics, how many are enrolled in both chemistry and physics?

A)   3
B)   6
C)   9
D)   12
E)   29

44. Standardized Test (ꝑ ꝑ)

If 85% of the test takers taking an old paper and pencil standardized exam answered the first question on a given math section correctly, and 75% of the test takers answered the second question correctly, and 65% of the test takers answered both questions correctly, what percent answered neither correctly?

A)   5%
B)   10%
C)   15%
D)   20%
E)   25%

45. Language Classes (🌶🌶🌶)

According to the admissions and records office of a major university, the schedules of X first-year college students were inspected and it was found that S number of students were taking a Spanish course, F number of students were taking a French course, and B number of students were taking both a Spanish and a French course. Which of the following expressions gives the percentage of students whose schedules were inspected who were taking neither a Spanish course nor a French course?

A) $100 \times \dfrac{X}{B + F + S}$

B) $100 \times \dfrac{B + F + S}{X}$

C) $100 \times \dfrac{X - F - S}{X}$

D) $100 \times \dfrac{X + B - F - S}{X}$

E) $100 \times \dfrac{X - B - F - S}{X}$

46. Valley High (🌶🌶🌶)

At Valley High School, four students have chosen to do an independent study project in biology, four in chemistry, and four in physics. Two students have chosen to do independent study projects in both biology and chemistry, two in both chemistry and physics, and two in both biology and physics. One student has chosen to do an independent study project in all three of biology, chemistry, and physics. How many students are involved in the above-mentioned independent study projects?

A) 5
B) 7
C) 9
D) 12
E) 19

47. German Cars (🐚🐚🐚)

The New Marketing Journal conducted a survey of wealthy German car owners. According to the survey, all wealthy car owners owned one or more of the following three brands: BMW, Mercedes, or Porsche. Respondents' answers were grouped as follows: 45 owned BMW cars, 38 owned Mercedes cars, and 27 owned Porsche cars. Of these, 15 owned both BMW and Mercedes cars, 12 owned both Mercedes and Porsche cars, 8 owned both BMW and Porsche cars, and 5 persons owned all three types of cars. How many different individuals were surveyed?

A) 70
B) 75
C) 80
D) 110
E) 130

Matrix Scenarios

☞ *Tip #15:* *A matrix is a marvelous tool for summarizing data that is being contrasted across two variables and that can be sorted into four distinct outcomes.*

Snapshot

A matrix is a box-like table. In fact, the simplest matrix is a table with exactly nine boxes—four boxes on the inside and five boxes on the outside. Our job with respect to matrix problems is to fill in known information and, through simple arithmetic, find the unknown information. A completed matrix will show data in all nine boxes.

Take, for example, a batch of 100 toys, fresh off the production line. Each toy has exactly two of four characteristics: it is either blue or green and either large or small. The total number of toys equals 100, and this figure appears in the bottom right corner of the extended matrix. The dotted lines are merely useful extensions for use in totaling data.

Factory Toys

In our batch of 100 toys, 55 are blue and the rest are green. Half the toys are large and half the toys are small. If 15 of the toys are small and green, how many are large and blue?

Here is the nine-box table used to set up this problem:

Color

		Blue	Green	
Size	Large	X	X	XX
	Small	X	X	XX
		XX	XX	XXX

Let's fill in the given numbers:

Color

		Blue	Green	
Size	Large	?		50
	Small		15	50
		55	45	100

Finally, we total down and across:

Color

		Blue	Green	
Size	Large	? = 20	30	50
	Small	35	15	50
		55	45	100

Why do matrixes work so neatly? Things work neatly as long as all data is mutually exclusive and collectively exhaustive. What does this mean? Mutually exclusive is a fancy way of saying that the data does not overlap; it is distinct. In other words, toys must be either blue or green and either large or small. We can't have toys which are both blue and green (e.g., blue-green or striped) or both large and small (i.e., medium-sized). Collectively exhaustive means that the number of data is finite. There are only 100 toys, of which exactly 55 are blue, 45 are green, 50 are large, and 50 are small. The fact that this data is mutually exclusive and collectively exhaustive allows everything to total both down and across.

Because matrixes handle information so neatly, it is not surprising that they are a consultant's favorite presentation tool. Legend has it that one young consultant became so enamored with matrixes that he called them "boxes of joy"!

48. Job Search (🌶)

Of 35 applicants applying for a job, 20 had at least 7 years' work experience, 23 had degrees, and 3 had less than 7 years' work experience and did not have a degree. How many of the applicants had at least 7 years' work experience and a degree?

A)   3
B)   9
C)   11
D)   12
E)   20

49. Nordic (🌶)

In a certain town in Norway, 75% of the people have blond hair, 65% have blue eyes, and 45% have both blond hair and blue eyes. What percentage of the people in the town has neither blond hair nor blue eyes?

A)   5%
B)   20%
C)   30%
D)   45%
E)   65%

50. Singles ( )

In a graduate physics course, 70% of the students are male and 30% of the students are married. If two-sevenths of the male students are married, what fraction of the female students is single?

A) $\dfrac{2}{7}$ B) $\dfrac{1}{3}$ C) $\dfrac{5}{7}$ D) $\dfrac{2}{3}$ E) $\dfrac{5}{7}$

51. Batteries ( )

One-fifth of the batteries produced by a certain upstart factory are defective, and one-quarter of all batteries produced are rejected by the quality control technician. If one-tenth of the non-defective batteries are rejected by mistake, and if all the batteries not rejected are sold, then what percentage of the batteries sold by the factory are defective?

A)   4%
B)   5%
C)   6%
D)   8%
E)   12%

52. Experiment (🌶🌶🌶)

Sixty percent of the rats included in an experiment were female. If some of the rats died during an experiment and 70% of the rats that died were male rats, what was the ratio of the death rate among the male rats to the death rate among the female rats?

A) 7 : 2
B) 7 : 3
C) 2 : 7
D) 3 : 7
E) Cannot be determined from the information given

Mixture Scenarios

☞ *Tip #16:* *Mixture problems are best solved using the barrel method, which enables information to be summarized in a 3-row by 3-column table.*

Snapshot

There are essentially two kinds of mixture problems: dry mixtures and wet mixtures. Dry mixtures involve dry ingredients, such as candies, nuts, or coffee beans, whereas wet mixtures involve liquids such as chemicals. The amount or quantity of a mixture is usually measured in liters, gallons, pounds, or kilograms. Below is a template, herein called the barrel method, for use in solving wet or dry mixture problems.

As illustrated in the next example, this technique works miraculously in helping to structure mixture problems. Example: How many liters of pure acid must be added to 40 liters of a 10% solution of acid to make a 25% acid solution?

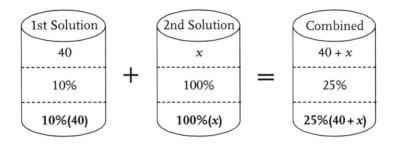

Next we create an equation from the totals (in bold) appearing at the bottom of the barrels:

$$10\%(40) + 100\%(x) = 25\%(40 + x)$$
$$4 + x = 10 + 0.25x$$
$$1.0x - 0.25x = 10 - 4$$
$$0.75x = 6$$
$$x = \frac{6}{0.75}$$
$$x = 8 \text{ liters}$$

53. Petroleum ( )

50 gallons of a 40% oil mixture is added to 25 gallons of a 20% oil mixture. What percentage of the resulting mixture is oil?

A)   25%

B)   30%

C) $33\frac{1}{3}\%$

D)   35%

E)   45%

54. Perfect Blend ( )

A dealer wishes to mix 20 kilograms of coffee selling for $7.50 per kilogram with some more expensive coffee selling for $15.00 per kilogram to make a mixture that will sell to coffee houses for $12.00 per kilogram. How many kilograms of the more expensive coffee should be used?

A)   25
B)   30
C)   35
D)   40
E)   45

55. Nuts ( )

A wholesaler plans to sell 100 pounds of mixed nuts at $2.50 a pound. She mixes peanuts worth $1.50 a pound with cashews worth $4.00 a pound. How many pounds of cashews does she use?

A)   40
B)   45
C)   50
D)   55
E)   60

56. Adding H_2O (🌶🌶)

How many gallons of water must be added to 40 gallons of a solution that is 30% salt to dilute it to a solution that is 25% salt?

A) 6
B) 8
C) 9
D) 10
E) 12

57. Evaporation (🌶🌶🌶)

How many liters of water must be evaporated from 50 liters of a 3% sugar solution to get a 10% sugar solution?

A) 15
B) $16\frac{2}{3}$
C) 27
D) $33\frac{1}{3}$
E) 35

58. Gold (🌶🌶🌶)

An alloy weighing 24 ounces is 70% gold. How many ounces of pure gold must be added to create an alloy that is 90% gold?

A) 6
B) 9
C) 12
D) 24
E) 48

Weighted Average Scenarios

☞ *Tip #17:*   *To find a weighted average, multiply each event by its associated weight and total the results. In the case of probabilities, multiply each event by its respective probability and total the results.*

Snapshot

The weighted average concept is actually quite intuitive. To find weighted average, we multiply "events" by their respective "weight" and total the results. Events are the things that we wish to rate, rank, or judge. Weights refer to the amount of emphasis we want to attribute to each event and are commonly expressed as percentages, fractions, decimals, or probabilities. The beauty of weighted average is that we can assign different weights based on the relative importance of events—the more important the event, the more weight it is given.

Below is the weighted average formula for two events:

$$WA = (Event_1 \times Weight_1) + (Event_2 \times Weight_2)$$

An alternative format, for ease of computation, unfolds as follows:

$$Event_1 \times Weight_1 = xx$$
$$Event_2 \times Weight_2 = \underline{xx}$$
$$\underline{\underline{xx}}$$

Exam Time

A student scores 60 out of 100 points on his midterm exam and 90 out of 100 points on his final exam. If the exams are both weighted equally, counting for 50% of the student's final course grade, then what is his course grade?

$$60 \times 50\% = 30$$
$$90 \times 50\% = \underline{45}$$
$$75$$

Based on the same information above, what is the student's final course grade if the midterm exam is weighted 40% and the final exam is weighted 60 percent?

$$60 \times 40\% = 24$$
$$90 \times 60\% = \underline{54}$$
$$78$$

Note that the weights above could also have been expressed in terms of fractions or decimals:

$$60 \times \frac{40}{100} = 24$$
$$90 \times \frac{60}{100} = \underline{54}$$
$$78$$

$$60 \times 0.4 = 24$$
$$90 \times 0.6 = \underline{54}$$
$$78$$

Hiring and promotion decisions are classic examples of situations in which subjective influences can override an objective decision-making process. Weighted ranking therefore presents a method to quantify decision opportunities.

Consider a company with 10 salespersons, one of whom is to be named National Sales Manager. The 10 candidates are first ranked from 1 to 10 (10 being the highest rating) across three criteria. The three criteria—technical skills, people skills, and track record—are weighted using the weights of 0.2, 0.3 and 0.5, respectively. Note that instead of using decimals (0.2, 0.3, 0.5), we could also use percentages (i.e., 20%, 30%, 50%), fractions (i.e., $\frac{2}{10}, \frac{3}{10}, \frac{5}{10}$), or even whole numbers, such as 2, 3, and 5.

Based on the results from weighting the data, Sabrina receives the highest ranking, while George gets the next highest ranking. The weights used will typically total to 1, as is so often the case when dealing with percentages, fractions, decimals, or probabilities. Sometimes problems will use arbitrary weights which do not equal to 1.

Chess

In chess, a pawn is worth 1 point, a knight or bishop is worth 3 points, a rook is worth 5 points, and a queen is worth 9 points. Player A has two rooks, a knight, and 3 pawns. Player B has a bishop, 4 pawns, and a queen. Who is ahead and by how much?

	Player A	Player B
Pawns:	$3 \times 1 = 3$ pts	$4 \times 1 = 3$ pts
Bishops:		$1 \times 3 = 3$ pts
Knights:	$1 \times 3 = 3$ pts	
Rooks:	$2 \times 5 = 10$ pts	
Queens:		$1 \times 9 = 9$ pts
	16 points	16 points

Answer: Both players are tied at 16 points each.

Exhibit 2.1   Performance of Salespersons

	Technical skills and product knowledge	People skills and communication ability	Track record and ability to get things done
Albert	2	3	7
Betty	5	2	6
George	4	5	9
Jed	3	7	1
Jono	7	10	3
Martha	10	1	2
Patricia	1	4	8
Randy	8	9	4
Sabrina	3	6	10
William	6	8	5

EXHIBIT 2.2   PERFORMANCE USING WEIGHTED AVERAGE

	Technical skills and product knowledge Weight = 0.2	People skills and communication ability Weight = 0.3	Track record and ability to get things done Weight = 0.5	Total
Albert	2 × 0.2 = 0.4	3 × 0.3 = 0.9	7 × 0.5 = 3.5	4.8
Betty	5 × 0.2 = 1.0	2 × 0.3 = 0.6	6 × 0.5 = 3.0	4.6
George	4 × 0.2 = 0.8	5 × 0..3 = 1.5	9 × 0.5 = 4.5	6.8
Jed	3 × 0.2 = 0.6	7 × 0.3 = 2.1	1 × 0.5 = 0.5	3.2
Jono	7 × 0.2 = 1.4	10 × 0.3 = 3.0	3 × 0.5 = 1.5	5.9
Martha	10 × 0.2 = 2.0	1 × 0.3 = 0.3	2 × 0.5 = 1.0	3.3
Patricia	1 × 0.2 = 0.2	4 × 0.3 = 1.2	8 × 0.5 = 4.0	5.4
Randy	8 × 0.2 = 1.6	9 × 0.3 = 2.7	4 × 0.5 = 2.0	6.3
Sabrina	3 × 0.2 = 0.6	6 × 0.3 = 1.8	10 × 0.5 = 5.0	7.4
William	6 × 0.2 = 1.2	8 × 0.3 = 2.4	5 × 0.5 = 2.5	6.1

EXHIBIT 2.3   RANKING OF SALESPERSONS

	Technical skills and product knowledge Weight = 0.2	People skills and communication ability Weight = 0.3	Track record and ability to get things done Weight = 0.5	Total votes (The higher the better)	Rank (The lower the better)
Sabrina	3 × 0.2 = 0.6	6 × 0.3 = 1.8	10 × 0.5 = 5.0	7.4	1
George	4 × 0.2 = 0.8	5 × 0..3 = 1.5	9 × 0.5 = 4.5	6.8	2
Randy	8 × 0.2 = 1.6	9 × 0.3 = 2.7	4 × 0.5 = 2.0	6.3	3
William	6 × 0.2 = 1.2	8 × 0.3 = 2.4	5 × 0.5 = 2.5	6.1	3
Juno	7 × 0.2 = 1.4	10 × 0.3 = 3.0	3 × 0.5 − 1.5	5.9	5
Patricia	1 × 0.2 = 0.2	4 × 0.3 = 1.2	8 × 0.5 = 4.0	5.4	6
Albert	2 × 0.2 = 0.4	3 × 0.3 = 0.9	7 × 0.5 = 3.5	4.8	7
Betty	5 × 0.2 = 1.0	2 × 0.3 = 0.6	6 × 0.5 = 3.0	4.6	8
Martha	10 × 0.2 = 2.0	1 × 0.3 = 0.3	2 × 0.5 = 1.0	3.3	9
Jed	3 × 0.2 = 0.6	7 × 0.3 = 2.1	1 × 0.5 = 0.5	3.2	10

Sweet Sixteen

On her 16th birthday, Jane received $500 from each of her two uncles. Both amounts had been deposited in two local banks, one bank paying 6 percent per annum and the other paying 7 percent per annum. How much in total did she earn from these two investments over the course of exactly one year?

$500 × 6% = $30
$500 × 7% = $35
             $65

59. Bowling (🌶)

In a bowling tournament, 60% of the bowlers had an average (arithmetic mean) score of 110, and the other 40% had an average score of 130. If each of the bowlers bowled an equal number of games, what is the average (arithmetic mean) score of all the bowlers in the tournament?

A) 115
B) 118
C) 120
D) 122
E) 125

60. The Right Mix ( )

Chocolate A costs $9 per kilogram and chocolate B costs $15 per kilogram. What is the average cost per kilogram of a combined mixture that is 80% chocolate A and 20% chocolate B?

A) $9.80
B) $10.00
C) $10.20
D) $12.00
E) $13.80

61. Baseball ( )

A baseball bat and a catcher's mitt currently have the same price. If the price of a baseball bat rises 3% and the price of a catcher's mitt rises 5%, how much more (in percentage terms) will it cost to buy 3 baseball bats and 3 catcher's mitts?

A) 4%
B) 8%
C) 16%
D) 24%
E) 48%

62. Venture Capital ( )

The probability that a certain new venture will generate a return of $5,000,000 is estimated to be 20%. The probability that the new venture will generate a return of $1,500,000 is 50%. The probability that the new venture will generate a return of $500,000 is 30%. Based on a weighted-average assessment, how much money can be expected to be earned?

A) $1,850,000
B) $1,900,000
C) $1,950,000
D) $2,000,000
E) $2,050,000

63. Portfolio (🌶🌶)

An investor is looking at three different investment possibilities. The first investment opportunity has a $\frac{1}{6}$ chance of returning $90,000, a $\frac{1}{2}$ chance of returning $50,000, and a $\frac{1}{3}$ chance of losing $60,000. A second investment opportunity has a $\frac{1}{2}$ chance of returning $100,000 and a $\frac{1}{2}$ chance of losing $50,000. The third investment opportunity has a $\frac{1}{4}$ chance of returning $100,000, a $\frac{1}{4}$ chance of returning $60,000, a $\frac{1}{4}$ chance of losing $40,000, and a $\frac{1}{4}$ chance of losing $80,000. Assuming the investor chooses to invest in all three investments, what will be his or her expected return?

A) $55,000
B) $60,000
C) $65,000
D) $70,000
E) $75,000

Chapter 3

Favorite Numeracy Dishes

There are three kinds of people in this world: those who can count and those who can't.
                              —Anonymous

Markup vs. Margin

🖝 *Tip #18:* *Markup is always larger than margin.*

Snapshot

The following story highlights the important real-life distinction between markup and margin:

A boss told his sales clerk to put price tags on all newly arrived toys and transfer them to the sales floor. The boss said, in mumbled language, that he wanted to adhere to company policy that requires earning a 25% return on all goods sold. However, the clerk heard "25%" and, looking at the cost of each toy item, promptly marked the goods up 25% based on the cost sheets. The company accountant eventually discovered this mistake when investigating why overall profit had fallen. The confusion created was the result of failing to distinguish between the terms markup and margin. Markup is always based on product cost, and gross margin is based on selling price. In short, markup is always a larger number than margin. Mathematically, markup is based on product cost—a smaller number than sales price—and consequently must be a larger percentage. Margin is based on selling price—a larger number—and consequently results in a smaller percentage. In other words, the markup required to earn a 25% gross margin is $33\frac{1}{3}\%$, not 25%. The clerk should have multiplied the cost of each good by 1.33, not 1.25.

Margin, also called "gross margin," is calculated as the difference between sales price and product cost. Markup is the difference between sales price and product cost. Margin and markup are commonly expressed as percentages. Thus, margin is calculated as the difference between sales price and product cost divided by sales price, which is then multiplied by 100. Markup is calculated as the difference between sales price and product cost divided by product cost, which is then multiplied by 100.

Four formulas are used when dealing with margin and markup. The first two are used when working with dollars (or other currencies) in order to calculate margin or markup percentages:

$$\text{Margin \%} = \frac{\text{Sales (\$)} - \text{Cost (\$)}}{\text{Sales (\$)}} \times 100\%$$

$$\text{Markup \%} = \frac{\text{Sales (\$)} - \text{Cost (\$)}}{\text{Cost (\$)}} \times 100\%$$

The next two formulas are used when working directly with percentages. Here's the formula for converting markup percentage to gross margin percentage:

$$\text{Margin \%} = \frac{\text{Markup \%}}{100\% + \text{Markup \%}}$$

This is the formula for converting gross margin percentage to markup percentage:

$$\text{Markup \%} = \frac{\text{GM \%}}{100\% - \text{GM \%}}$$

Multipliers

If we want to achieve a margin of 16.67%, then we mark up goods by 20%. Or put another way, if goods are marked up by 20%, then a margin of 16.67% will be achieved. Of course, the easy way to arrive at this percentage is to use the "multipliers" given in exhibit 3.1. To achieve a gross margin of 16.67%, we simply multiply product cost by 1.20.

$$\text{Markup \%} = \frac{\text{GM \%}}{100\% - \text{GM \%}}$$

$$\text{Markup}\% = \frac{16.67\%}{100\% - 16.67\%} = \frac{16.67\%}{83.33\%} = 20\%$$

$$\text{Margin}\% = \frac{\text{Markup}\%}{100\% + \text{Markup}\%}$$

$$\text{Margin}\% = \frac{20\%}{100\% + 20\%} = \frac{20\%}{120\%} = 16.67\%$$

EXHIBIT 3.1 COMMON GROSS MARGIN AND MARKUP PERCENTAGES

Gross Margin	Markup	Multipliers – Multiply product cost by these factors to achieve gross margin percentages
16.67%	20%	1.20
20%	25%	1.25
25%	$33\frac{1}{3}\%$	1.33
$33\frac{1}{3}\%$	50%	1.50
50%	100%	2.00

64. Gross Margin ( )

If a company buys a product for $7.50 and sells the product for $10, what is its gross margin percentage?

A) 20%
B) 25%
C) $33\frac{1}{3}\%$
D) 50%
E) 100%

65. Markup (🌶)

If a company buys a product for $7.50 and sells the product for $10, what is its markup percentage?

A) 20%
B) 25%
C) $33\frac{1}{3}$%
D) 50%
E) 100%

66 Alpha Inc. (🌶🌶)

Alpha Inc. habitually marks its products up 50%. How much is the corporation making on each product sold with respect to a gross margin percentage?

A) 20%
B) 25%
C) $33\frac{1}{3}$%
D) 50%
E) 100%

67. Beta Corp. (🌶🌶)

Beta Corp. wants to sell its products and achieve a gross margin of 20%. What should the markup percentage be to achieve this gross margin percentage?

A) 20%
B) 25%
C) $33\frac{1}{3}$%
D) 50%
E) 100%

Price, Cost, Volume, and Profit

☞ *Tip #19: An increase (decrease) in sales volume does not necessarily translate to an increase (decrease) in profit.*

Snapshot

Price-cost-volume-profit scenarios highlight the interplay of the three profit components: price, cost, and volume. Just because a computer store is selling more computers than last year does not mean that its profits are up. Why? Because an increase (decrease) in sales volume (i.e., number of unit sales) of a good or service item does not necessarily equal an increase (decrease) in revenues or profits. Likewise, an increase (decrease) in the price of a good or service does not necessarily equal an increase (decrease) in profits. Moreover, an increase (or decrease) in the cost of a good or service does not necessarily translate to a decrease (or increase) in profits.

Note that the words "expense" and "cost" are deemed synonymous; the words "volume" and "number of units" are interchangeable.

Here are three pertinent formulas:

Cost Formula

$$\text{Cost}_{\text{per unit}} \times \text{Number of units} = \text{Cost}$$

$7 per unit × 1,000 units = $7,000

Revenue Formula

$$\text{Price}_{\text{per unit}} \times \text{Number of units} = \text{Revenue}$$

$10 per unit × 1,000 units = $10,000

Profit Formula

Revenue − Cost = Profit

$10,000 − $7,000 = $3,000

Two variations of profit formulas include:

$(Price_{per\ unit} − Cost_{per\ unit}) × No.\ of\ units = Profit$

($10 per unit − $7 per unit) × 1,000 units = $3,000

$(Price_{per\ unit} × No.\ of\ units) − (Cost_{per\ unit} × No.\ of\ units) = Profit$

($10 per unit × 1,000 units) − ($7 per unit × 1,000 units) = $3,000

68. Century Bus Line (🌶)

The Century Bus Line's financial forecast for the coming year indicates a decrease in overall operating profits. However, the company forecasts a record number of paying passengers for the coming year. Which of the following facts, A through E below, would best contribute to an explanation of this scenario?

A) The average length of a Century Bus Line trip will decrease while the average number of trips will increase.

B) There is to be no projected increase in the number of canceled trips in the coming year.

C) Two new competitors of Century Bus Line are expected to enter the market in the coming year.

D) Century Bus Line will cut ticket prices by an average of 25% in the coming year.

E) Century Bus Line will be able to reduce significantly its overall operating costs due to government deregulation of fuel in the coming year.

69. High-Speed Boat (🌶🌶)

To operate an ocean high-speed boat, it costs g cents a mile for gasoline and h cents a mile for oil. How many <u>dollars</u> will it cost, in terms of gasoline and oil, to run the high-speed boat for 100 miles?

A) g
B) $g + h$
C) $g + 0.1h$
D) $100g + 100h$
E) $(g + h)/100$

70. Cases (🌶🌶)

A manufacturer of cassette cases wants to make a profit of x dollars. When it sells 10,000 cassette cases it costs 5 cents (=$0.05) to make the first 1,000 cases, and then it costs y cents a case to make the remaining 9,000 cassette cases. What price in dollars should it charge for the 10,000 cassette cases?

A) $50 + 10y + x$
B) $50 + 90y + x$
C) $10,000 + 1,000y$
D) $10,000 + 9.000y$
E) $10,000 + 1,000y + 100x$

71. Delicatessen (🌶🌶)

A large delicatessen purchased p pounds of cheese for c dollars per pound. If d pounds of the cheese had to be discarded due to spoilage and the delicatessen sold the rest for s dollars per pound, which of the following represents the gross profit on the sale of the purchase (gross profit equals sales revenue minus product cost)?

A) $(p - d)(s - c)$
B) $s(p - d) - pc$
C) $c(p - d) - ds$
D) $d(s - c) - pc$
E) $pc - ds$

🖙 *Tip #20:*   *When price-cost-volume-profit problems are presented in the form of algebraic expressions, it is best to first find a dollar per unit, dollar per person, usage per unit, or usage per person figure.*

Snapshot

The statement, "c pounds of carrots cost 20 dollars" results in a dollar per unit figure expressed as 20 dollars/c pounds or 20/c dollars per pound. Note that in these problems the dollar figure becomes the fraction's numerator, not its denominator. Using the cost formula, we solve the problem below as follows:

$$\text{Cost} = \text{Cost}_{\text{per unit}} \times \text{Number of units}$$

$$\text{Cost} = \frac{20 \text{ dollars}}{c \text{ \cancel{pounds}}} \times N \text{ \cancel{pounds}}$$

$$\text{Cost} = \frac{20N}{c} \text{ dollars}$$

72. Paper Clips ( )

If p paper clips cost c cents, how many cents do q paper clips cost?

A) $\dfrac{cq}{p}$ B) $\dfrac{pq}{c}$ C) $\dfrac{p}{cq}$ D) $\dfrac{c}{pq}$ E) pcq

73. Garments ( )

If s shirts can be purchased for d dollars, how many shirts can be purchased for t dollars?

A) sdt B) $\dfrac{ts}{d}$ C) $\dfrac{td}{s}$ D) $\dfrac{d}{st}$ E) $\dfrac{s}{dt}$

74. Copy ( )

If each photocopy of a manuscript costs 2 cents per page, what is the cost, in cents, to reproduce y copies of a y-page manuscript?

A) $2y$
B) y^2
C) $2y^2$
D) $4y$
E) $4y^2$

75. Pete's Pet Shop (🐦🐦)

At Pete's Pet Shop, 35 cups of bird seed are used every 7 days to feed 15 parakeets. How many cups of bird seed would be required to feed 9 parakeets for 12 days?

A) 32
B) 36
C) 39
D) 42
E) 45

Break-Even Point

☞ Tip #21: *In break-even point problems, break-even occurs exactly where variable revenue (sales revenue less variable costs) equals total fixed costs.*

Snapshot

Break-even point is defined as the point where zero profit is achieved. That is, the point where sales revenue exactly equals total costs. Total cost is divided into fixed and variable components because this facilitates break-even point calculations.

Variable costs vary directly with units produced, whereas fixed costs do not. Fixed costs therefore stay the same regardless of whether you sell one item or a million items.

In everyday living, rent is considered a fixed cost, whereas utilities (water, gas, and electricity usage) are typically considered variable costs. In terms of rent, it does not matter how much you use your apartment in one month—the cost is the same. In terms of utilities, the more you use in one month, the more you pay.

Suppose that company A produces widget Z. Assume that a company produces only widget Zs and has fixed costs of $4,500, a selling price of $10 per unit, and a variable cost of $2.50 per unit. What is its break-even point in *units*?

$$\text{BE}_{\text{units}} = \frac{\text{Fixed Costs}}{\text{Selling Price}_{\text{per unit}} - \text{Variable Cost}_{\text{per unit}}}$$

$$\text{Break Even}_{\text{units}} = \frac{\$4,500}{\$10 - \$2.50}$$

$$\text{Break Even}_{\text{units}} = \frac{\$4,500}{\$7.50} = 600 \text{ units}$$

What if we wanted instead to determine the break-even point in terms of sales *dollars*?

$$\text{Break Even}_{\text{dollars}} = \frac{\text{Fixed Costs}}{\dfrac{\text{SP}_{\text{per unit}} - \text{VC}_{\text{per unit}}}{\text{SP}_{\text{per unit}}}}$$

$$\text{Break Even}_{\text{dollars}} = \frac{\$4,500}{\dfrac{\$10 - \$2.50}{\$10}}$$

$$\text{Break Even}_{\text{dollars}} = \frac{\$4,500}{\dfrac{\$7.50}{\$10}}$$

$$\text{Break Even}_{\text{dollars}} = \frac{\$4,500}{0.75} = \$6,000$$

Of course, we can prove all of this by remembering that break-even is that point where total revenues equal total costs.

Total Revenue = Total Costs
Total Revenue = Fixed Costs + Variable Costs

$600 \text{ units} \times \$10 \text{ per unit} = \$4{,}500 + (600 \text{ units} \times \$2.50 \text{ per unit})$

$\$6{,}000 = \$4{,}500 + \$1{,}500$

$\$6{,}000 = \$6{,}000$

Break-even point theory is sometimes applied to competing projects. Different projects invariably have different cost and revenue structures. When two projects are being evaluated on competing terms, we may want to know how many additional sales are needed before the costs, profits, or revenues of one project become equal to the other.

Set the two projects "equal" to one another. Fill in given information, and solve for the remaining variable.

$$\text{Revenue}_{\text{Project 1}} = \text{Revenue}_{\text{Project 2}}$$

$$\text{Price}_{\text{per unit}} \times \text{No. of units} = \text{Price}_{\text{per unit}} \times \text{No. of units}$$

$$\text{Cost}_{\text{Project 1}} = \text{Cost}_{\text{Project 2}}$$

$$\text{Cost}_{\text{per unit}} \times \text{No. of units} = \text{Cost}_{\text{per unit}} \times \text{No. of units}$$

$$\text{Profit}_{\text{Project 1}} = \text{Profit}_{\text{Project 2}}$$

$$\text{Profit}_{\text{per unit}} \times \text{No. of units} = \text{Profit}_{\text{per unit}} \times \text{No. of units}$$

76. Book Publisher (🌶🌶)

The chart below lists the price and cost data for publication of a soon-to-be released paperback novel. What is the minimum number of copies that must be sold in order for the publisher not to lose money on the book?

Bookstore Price	$5.95/copy
Fixed Costs	
Author fees (advance)	$150,000
Editing & Design	$20,000
Advertising & Promotion	$230,000
Variable Costs	
Paper, binding and printing	33¢/copy
Storage and distribution	12¢/copy
Author's royalties	50¢/copy

A) 20,000
B) 48,358
C) 50,000
D) 67,143
E) 80,000

77. Sabrina to Change Jobs (🌶🌶🌶)

Sabrina is contemplating a job switch. She is thinking of leaving her job paying $85,000 per year to accept a sales job paying $45,000 per year plus 15% commission for each sale made. If each of her sales is for $1,500, what is the smallest number of sales she must make per year if she is not to lose money because of the job change?

A) 26
B) 177
C) 178
D) 200
E) 600

Aggregate Costs vs. Per-Unit Costs

☞ *Tip #22:* *Total costs must be distinguished from per-unit costs, as is the case whenever the number of units varies between two situations under comparison.*

Snapshot

A distinction must be drawn between aggregate (or total costs) and per-unit costs because total costs often are compared without regard to the number of units involved. We have to be careful not to assume that the total number of units is constant. For example, a single piece of Dove chocolates and a single piece of Valentine chocolates are of equivalent size and of comparable quality. Since a box of Dove chocolates cost £5 (British pounds) and a box of Valentine chocolates costs £4 (British pounds), can we conclude Valentine chocolates are our best buy? Not necessarily—what if a box of Dove chocolates contains significantly more chocolates compared with a box of Valentine chocolates?

When making per-unit comparisons between two scenarios, it is important to visualize the three components of the cost formula:

Scenario 1   *Scenario 2*

$$\frac{\text{Total Costs}}{\text{Units}} = \text{Per unit cost} \qquad \frac{\text{Total Costs}}{\text{Units}} = \text{Per unit cost}$$

78. Grapes (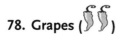)

It costs much less to grow an acre of grapes in California than to grow an acre of watermelons in Oklahoma. This fact should be obvious to anyone who read last year's annual farm report, which clearly shows that many millions of dollars more were spent last year growing watermelons in Oklahoma than were spent growing grapes in California.

Which of the following, if true, would most seriously call into question the reasoning above?

A) Profits on the sale of California grapes are much higher than the profits derived from the sale of Oklahoma watermelons.

B) Far fewer total acres of grapes were grown in California last year than total acres of watermelons in Oklahoma.

C) An acre of grapes in Oklahoma costs much more to grow than an acre of watermelons in California.

D) Part of the California-grown grapes were used to feed livestock, whereas all of the watermelons grown in Oklahoma were used for human consumption.

E) State subsidies accounted for a larger percentage of the amount spent growing watermelons in Oklahoma than of the amount spent growing grapes in California last year.

Efficiency

🖝 *Tip #23:* *Efficiency is output divided by input. Cost efficiency is cost divided by efficiency.*

Snapshot

Although efficiency can be defined in a few different ways, perhaps the most multi-purpose definition appears below. Input is usually measured in units.

$$\text{Efficiency} = \frac{\text{Output}}{\text{Input}}$$

$$\text{Cost Efficiency} = \frac{\text{Cost}}{\text{Efficiency}} = \frac{\text{Cost}}{\dfrac{\text{Output}}{\text{Input}}}$$

Suppose that there are two employees working at a garment company. Employee A can make 10 garments in 5 hours. Employee B can make 5 garments in 5 hours. Their efficiencies become:

Employee A

$$\text{Efficiency} = \frac{\text{Output}}{\text{Input}} = \frac{10\,\text{garments}}{5\,\text{hours}} = 2\,\text{garments per hour}$$

Employee B

$$\text{Efficiency} = \frac{\text{Output}}{\text{Input}} = \frac{5\,\text{garments}}{5\,\text{hours}} = 1\,\text{garment per hour}$$

We can conclude that employee A is twice as efficient as employee B. But is employee A more cost efficient than employee B? Let's say that employee A is paid \$40 per hour but employee B is paid \$10 per hour.

Employee A

$$\text{Cost Efficiency} = \frac{\$40/\text{hour}}{\dfrac{10 \text{ garments}}{5 \text{ hours}}} = \$20 \text{ per garment}$$

Employee B

$$\text{Cost Efficiency} = \frac{\$10/\text{hour}}{\dfrac{5 \text{ garments}}{5 \text{ hours}}} = \$10 \text{ per garment}$$

So we can conclude that although employee A is twice as efficient as employee B, employee B is twice as cost-effective as employee A. Of course, we did make some assumptions. For example, we assumed that the quality of the craftsmanship is comparable between the two employees and that the slower employee B did not incur any additional costs when taking more hours to finish his or her work (i.e., factory overhead).

79. Act-Fast ( )

One Act-Fast tablet contains twice the pain reliever found in a tablet of regular aspirin. A consumer will have to take two aspirin to get the relief provided by one Act-Fast tablet. And since a bottle of Act-Fast costs the same as a bottle of regular aspirin, consumers can be expected to switch to Act-Fast.

Which of the following, if true, would most weaken the argument that consumers will be discontinuing the use of regular aspirin and switching to Act-Fast?

A)   A regular aspirin tablet is less than one-half as large as an Act-Fast tablet.

B)   Neither regular aspirin nor Act-Fast is as effective in relief of serious pain as are drugs available only by prescription.

C)   Some headache sufferers experience a brief period of nausea shortly after taking Act-Fast but not after taking regular aspirin.

D)   A regular bottle of aspirin contains more than twice as many tablets as does a bottle of Act-Fast.

E)   The pain reliever in Act-Fast is essentially the same pain reliever found in regular aspirin.

80. Prototype (𝄢 𝄢 𝄢)

A prototype fuel-efficient car (P-Car) is estimated to get 80% more miles per gallon of gasoline than does a traditional fuel-efficient car (T-Car). However, the P-Car requires a special type of gasoline that costs 20% more per gallon than does the gasoline used by a T-Car. If the two cars are driven the same distance, what percent less than the money spent on gasoline for the T-Car is the money spent on gasoline for the P-Car?

A)   16%

B) $33\frac{1}{3}\%$

C)   50%

D)   60%

E) $66\frac{2}{3}\%$

Distribution and Allocation

☞ *Tip #24:*   *Elements within distributions may be uniformly spaced or concentrated in "bunches." Allocations may be equal or unequal, proportionate or disproportionate.*

Snapshot

If we have 75 pencils to be distributed among 15 students, we will likely assume that each student will receive 5 pencils each. If a house has 3 rooms covering an aggregate area of 750 square feet, it is highly unlikely we will assume that each room is exactly 250 square feet. If a bookstore sells 5 dictionaries in 5 days, we may or may not be tempted to assume that a single book was sold each day. Perhaps multiple books were sold on one or more days while none were sold on other days. In short, distributions are not always proportional or linear.

81. Wanda ( )

Wanda is working on a take-home math test which contains 4 sections and a total of 100 true-false problems. If each problem is worth 1 point, and Wanda has finished the first 3 sections as well as the first problem of the fourth section, has she finished more than 75% of the problems on her test?

A)   Yes
B)   No
C)   Can't Tell

82. Noble Book Club ( )

A sales representative from the Noble Book Club recently hailed the club's free CD-ROM gift program as a big boost for sales. A year ago, the club began offering a free CD-ROM gift for any member ordering four or more books in a single month. Since then, the number of members ordering four or more books has risen by nearly 40%.

Which one of the following, if true, would most seriously weaken the representative's assessment of the program's effect on sales?

A)   The number of members ordering fewer than four books in a single month also rose in the last year.

B)   Most members ordered the same number of books over the year but concentrated their orders in specific months.

C)   The cost of providing free CD-ROM gifts nearly offset the increased revenue from higher sales.

D)   Most competing book clubs, many of which also sponsored free gift programs, saw a drop in sales last year.

E)   The membership in book clubs across the country rose by more than 15% over the past year.

☞ *Tip #25:*   *Front-loading occurs when a greater proportion of inputs (i.e., time, materials, costs, efforts, benefits) are entered early in a process. Back-loading occurs when a greater proportion of inputs are entered late in a process.*

Snapshot

Examples of front-loading and back-loading pervade real-life settings. If we use all of our vacation time early in the work year, we have front-loaded our vacation time and have no days left. If we save all our vacation time for the end of the year, our decision is to back-load our vacation days.

Another example stems from scholastic study habits. Each of us can recall, with some humor, a time when we crammed insanely for a final exam. If we plotted our study time over the course of an entire semester or quarter for a given course, a huge spike would indicate a disproportional amount of study time recorded late in the school term. This exemplifies back-loaded study, where the inputs of time and effort occur late in the process.

Now reflect on the traditional formal education process in developed countries. A person completes his or her formal schooling in a very concentrated manner, regardless of whether it ends with high school, community college, or university. There is typically little formal schooling after that. The assumption is that the kinds of skills that are learned in school (or are supposed to be learned) are to benefit a person through his or her entire life—certainly basic skills such as reading, writing, arithmetic, and reasoning. After all, formal study ends somewhere between the ages of 18 and 25, and yet the average life expectancy of a person living in developed countries is around 80 years of age. Formal education may be viewed as a front-loaded endeavor insofar as the inputs of time and effort occur early in the process.

83. Max Motorcycles (🌶🌶)

An advertisement designed to convince readers of the great durability of motorcycles manufactured by the Max Motorcycle Company cites as evidence the fact that half of all motorcycles built by the company in the last 10 years are still on the road today compared to no more than a third for any other manufacturer.

Which of the following, if true, most strongly supports the advertisement's argument?

A) After taking inflation into account, a new Max Motorcycle costs only slightly more than a new model did 10 years ago.

B) In the past 10 years, Max Motorcycles has made fewer changes in the motorcycles it manufactures compared with changes made by other motorcycle companies.

C) Owners of Max Motorcycles typically keep their motorcycles well maintained.

D) The number of motorcycles built by Max Motorcycle each year has not increased sharply in the past 10 years.

E) Max Motorcycles have been selling at relatively stable prices in recent years.

Chapter 4

Special Math Garnishments

Math is like love. It starts out simple, then it gets complicated.
—Anonymous

Translating Basic Graphs

☞ *Tip #26:* *Forward-slashing graphs indicate that two variables are directly related. Back-slashing graphs indicate that two variables are indirectly (inversely) related. Graphs formed with straight lines have variables which are linearly related. Graphs formed with curved lines have variables which are non-linearly related.*

Snapshot

All of the upcoming nine line graphs have two basic characteristics: (1) they are all either flat (horizontal) or sloping; and (2) they are composed of straight lines or curves.

All the graphs in problem 84 showcase two variables—money and time. Money appears on the vertical or y-axis, and time appears on the horizontal or x-axis. Graphs which are flat represent graphs with variables that do not vary with one another. On the other hand, graphs which are sloping represent variables that do vary with one another. If a graph slopes upward (whether a straight line or a curve), the two variables are said to be directly related to one another. This means that as one variable increases the other also increases. Conversely, as one variable decreases, the other deceases. Graphs that slope downward represent variables that are indirectly related to one another. This means that as one variable increases, the other decreases. Conversely, as one variable decreases the other increases.

Finally, graphs which are straight lines have variables which are linearly related. "Linear" means that the variables increase or decrease proportionally. Graphs which are curved lines have variables which are non-linearly related. "Non-linear" means that the variables increase or decrease disproportionately.

Graphs composed of *straight lines sloping upward* depict variables which are both directly and linearly related to one another. For every unit of increase in one variable, there is a proportional unit of increase in the other variable. A good real-life example might involve the painting of a room. For every hour you spend painting, you get one more unit of work done.

Graphs composed of *curved lines sloping upward* depict variables which are directly but non-linearly related to one another. For every unit of increase in one variable, there is more than one unit of increase in the other variable. A good real-life example might involve completing a difficult puzzle with many pieces. For each additional hour we spend working on it, we complete a disproportionate amount of the puzzle because the pieces start to fit.

Graphs composed of *straight lines sloping downward* depict variables which are indirectly but linearly related to one another. For every unit of increase in one variable, there is a proportional unit of decrease in the other variable. For every hour you spend painting your room, you get one less hour of leisure time.

Graphs composed of *curve lines sloping downward* depict variables which are both indirectly and non-linearly related to one another. For every unit of increase in one variable, there is a disproportionately larger decrease in the other variable. A good real-life example might involve study time and retention over a sustained study period. For every hour spent studying, we gain a unit of benefit. But after two or three hours, we experience diminishing returns for our time and effort spent.

84. Time & Money (🌶)

Use the following nine graphs in the exercise that follows. The topic is one that is dear to us all—salaries and wages. A one-chili rating is assigned to this problem as you only need to get six out of twelve questions correct in order to get the whole problem right!

Exhibit 2.1   Salespersons Performance

Rising Graphs (A,B,C,D)
$ = wages/salary

Flat Graphs (E,F,G,H)
$ = wage/salary

Falling Graphs (I,J,K,L)
$ = wage/salary

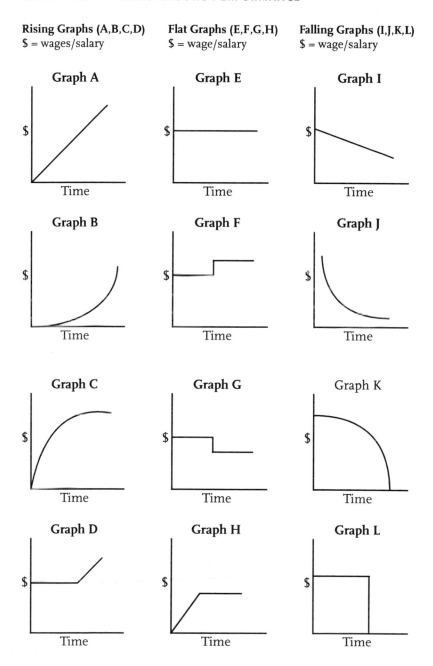

Choose one of the graphs appearing on the previous page and match it with one of the four statements below (within each of the four categories) by writing a letter in the box following each statement. The topic again is salaries and wages.

Rising Graphs (select from graphs A, B, C or D):

For every additional hour you work, you get paid incrementally more dollars for each hour worked. ❏

You get a fixed base salary but may also get additional pay based on hours worked which are not covered by your base salary. ❏

For every additional hour you work, you get paid incrementally fewer dollars for each hour worked. ❏

For every hour you work, you get paid y dollars per hour. ❏

Flat Graphs (select from graphs E, F, G or H):

You got a promotion at work which increased the fixed salary you now receive. ❏

Because of layoffs at work, you decided to take a reduction in your fixed monthly salary in order to keep your current job. ❏

You get paid a fixed (flat) wage regardless of the hours you work. ❏

You get paid by the hour, but cannot earn more than a certain dollar amount during the course of your working week. ❏

Falling Graphs (select from graphs I, J, K or L):

Your yearly pension decreases in terms of real dollars during each year of your retirement. ❏

Your wages decrease gradually with age until the time you stop working altogether. ❏

Your wages are fixed but stop the moment you retire. ❏

Your wages decrease dramatically with age, but you can still earn a little in your retirement. ❏

Line Graphs, Pie Charts, and Bar Charts

☞ *Tip #27:* *Line graphs are best used to show how data changes over time. Pie charts are great for breaking data down by category. Bar charts are ideal for side-by-side comparisons of up to six pieces of data. Knowing the strength of each type of graph also helps us interpret the underlying data.*

Snapshot

Three basic types of graphs and charts are commonly used to present data. The following questions would precipitate the use of each of these graphs or charts in the commercial realm:

Line graphs

"How have corporate profits, sales, or earnings per share increased or decreased over time?" Perhaps a line graph could show all three trends within a single graph.

Pie charts

"How are corporate expenses broken down by major categories?" For example, what percentages of total corporate expenses are administrative, salaries and wages, advertising, operating, and miscellaneous?

Bar charts

"How do the different corporate divisions stack up with one another in terms of head count (total number of employees)?"

How does our company rank alongside its major competitors in terms of market share?

Here's how these three types of graphs might be used in a non-work setting:

Line graphs

"Am I watching too much television?" After charting a week-by-week record of hours of TV viewing at home, you use the data to create a line graph showing the number of TV-free hours (at home) per week.

Pie charts

"Where is all my money going?" Use a pie chart to display what percentage of your disposable income is spent on different categories.

Bar charts

"What are my greatest personal strengths and how do they rank?" After identifying a half-dozen of your most salient personal or professional strengths, you assign values to them from 1 to 10 (1 being lowest and 10 being the highest) and display them using a bar chart.

Use the following two graphs to answer questions 85–87 regarding Grand Hotel & Casino's revenues for the month of January.

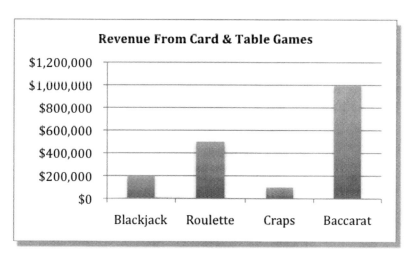

85. Grand Hotel Q1 (🌶🌶)

During the month of January, revenue from blackjack provided what percent of the total revenue for Grand Hotel & Casino?

86. Grand Hotel Q2 (🌶🌶)

How much money was earned through the sale of in-house shows?

87. Grand Hotel Q3 (🌶)

If the revenue from the rental of penthouse suites was equal to the revenue from miscellaneous items, what percent of the revenue from total room rental & restaurant sales was accounted for by penthouse suite rentals?

Use the following two graphs to answer questions 88–90 regarding the composition of Argentina's workforce between 1970 and 2009.

88. Workforce Q1 (🌶)

The decrease in blue-collar workers from 45% in 1970 to 20% in 2009 represents a decrease of how many millions of people?

89. Workforce Q2 (🌶🌶)

According to the data presented, the same number of people worked in the service sector in 2009 as worked in which sector in 1970?

90. Workforce Q3 (🌶🌶🌶)

From 1970 to 2009, the number of Argentinians working in the service, professional, and clerical sectors increased. Calculate the percentage increases for each of these sectors in terms of the number of workers. Also, calculate the percentage decrease for the blue-collar sector in terms of the number of workers. Which sectors, if any, experienced a percentage increase that was greater than the percentage decrease attributed to the blue-collar sector?

Use the following three graphs to answer questions 91–92 regarding Magna Fund's performance for the past four quarters.

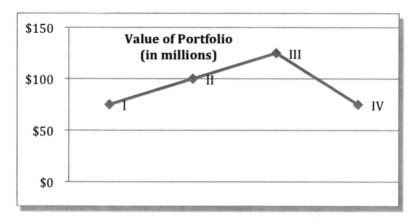

91. Magna Fund Q1 (🌶)

What was the percentage increase in the portfolio's value from the first quarter to the third quarter?

A) 40%
B) 50%
C) 66%
D) 125%
E) 166%

92. Magna Fund Q2 (🌶🌶🌶)

Which of the following statements can be inferred from the three graphs above?

I. The percent of the portfolio accounted for by cash declined from the third to the fourth quarter.

II. Bonds accounted for one-third or more of the portfolio in three of four quarters.

III. The average amount of dollar investment per individual investor was greatest during the fourth quarter.

IV. If the stock portfolio consisted of common and preferred shares only, and was maintained in a ratio of 70% common stock to 30% preferred stock, then the value of the preferred stock holdings was approximately $18 million during the second quarter.

A) I & II
B) II & III
C) III & IV
D) II, III & IV
E) All of the above

Correlation

☞ Tip #28: *Positive correlations have graphs sloping upward; negative correlations have graphs sloping downward. The stronger the correlation between two variables, the nearer to a straight line will be their graph; the graph of a perfect correlation is a straight line.*

Snapshot

Correlation is about the relationship between two (or more) variables. By relationship we are referring not only to how they are connected but also to the strength of their connection. Two variables can be connected in two basic ways: positively and negatively. Positively correlated variables are directly related. Negatively correlated variables are indirectly related.

The strength of the connection between two variables is measured by a value ranging from +1 to −1. A correlation of +1 is a positive-perfect correlation; a correlation of −1 is a negative-perfect correlation. In real life, positive-perfect or negative-perfect correlations are rare. Note that although the scatter points of perfect correlations appear as straight lines, positive-high correlations look like fat fountain pens, and positive-low correlations are shaped like squashes. Exhibit 4.1 provides a flowchart to summarize the statistical relationships as they relate to correlation. Exhibit 4.2 illustrates the different types of correlation using scatter plot graphs, which plot the intersection of data points.

EXHIBIT 4.1   CORRELATION FLOWCHART

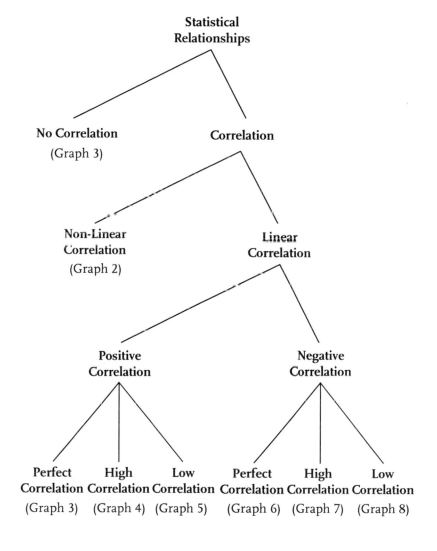

Graph 1 indicates no correlation. "What is the relationship between hair color and longevity?" We would expect the scatter points describing such data to be completely random.

Graph 2 depicts a non-linear correlation. "How is performance affected by anxiety levels?" Whether it be an academic or athletic undertaking, a moderate level of anxiety is optimal, with low or high levels of anxiety suboptimal.

Graph 3 is a positive-perfect correlation. The amount of coffee beans used and the number of cups of coffee consumed are directly related. Note that it would not be accurate to substitute the phrase "amount of coffee beans grown or produced" for "amount of coffee beans used" because coffee crop production might fluctuate independently of coffee beans actually consumed.

Graph 4 is a positive-high correlation. "How are company sales affected by increasing dollars spent on advertising?" We expect to see a very strong correlation between these two variables. The more we advertise, the more sales increase (within certain limits). In the case of advertising and sales, we would typically expect to see a 0.8 positive correlation.

Graph 5 is a positive-low correlation. "What is the relationship between height and weight?" The taller we are, the more we will likely weigh, but the correlation between height and weight is not particularly strong.

Graph 6 is a negative-perfect correlation. "How does the life experience we gain with each year's passing vary with respect to the number of years we have left to live?" Well, for every year we live and gain experience, we have one fewer year to live. It's a trade-off: one year of experience traded for one year of our life.

Graph 7 is a negative-high correlation. "How is our memory after drinking lots of beer?" It is a pretty good bet that for any individual, the more he or she drinks, the less remarkable will be his or her memory.

Graph 8 is a negative-low correlation. "How is one's level of physical fitness linked to food consumption?" There might be some truth to the idea that the more we eat the less fit we are. But we also know of cases where highly fit individuals eat copious amounts of food, as is the case with many professional athletes, particularly those playing contact sports such as American football or rugby.

EXHIBIT 4.2   CORRELATION ANALYSIS

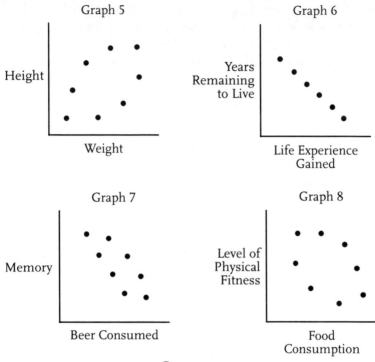

Graph 5

Height

Weight

Graph 6

Years Remaining to Live

Life Experience Gained

Graph 7

Memory

Beer Consumed

Graph 8

Level of Physical Fitness

Food Consumption

93. Correlation Exercise ( )

Placing the appropriate letter in each box, match graphs A through H, on the following page, with the caption below that best describes the type of correlation depicted.

Positive-perfect correlation ☐   Negative-perfect correlation ☐

Positive-high correlation ☐   Negative-high correlation ☐

Positive-low correlation ☐   Negative-low correlation ☐

Non-linear correlation ☐   Zero correlation ☐

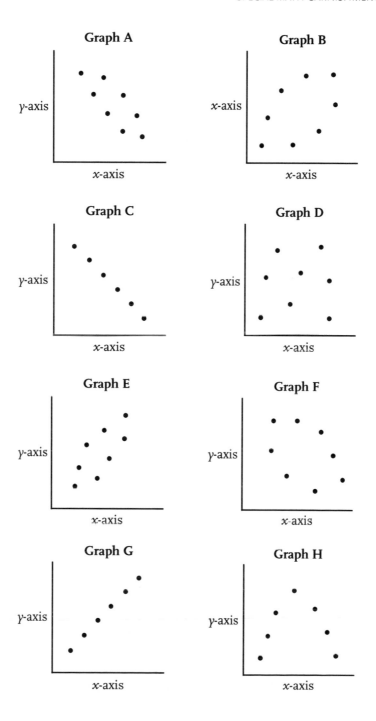

Measures of Average and Dispersion

☞ Tip #29:   *The mean, median, and mode for a given set of data may all be different values. Data with a high standard deviation is more spread out, and the graph of such data is typically "low" and "wide." Data with a low standard deviation is bunched, and its graph is typically "narrow" and "tall."*

Snapshot

Mean, median, and mode are referred to in statistics as measures of central tendency. They are ways of calculating averages. Mean (also called arithmetic mean or arithmetic average) is calculated as: average = sum ÷ number. We normally calculate average in this way for everyday purposes. Median is the middle-most number, found by taking half the numbers below and half the numbers above, after lining the numbers up in numerical sequence, from lowest to highest (or highest to lowest). Mode is the most frequently recurring number in a set of data.

Let's test the three concepts of central tendency—mean, median, and mode—using a simple example. A science teacher gives his or her class five pop quizzes during the semester, and a given student receives scores (out of a possible 10 points) of: 4.0, 4.0, 5.0, 7.0, and 10.0. What is this student's average score? Well it depends on which average you are referring to. In this case, the student would have an arithmetic average (mean) of 6.0, a median score of 5.0, and a mode of 4.0. On the other hand, if a student had quiz results of 5.0, 6.0, 9.0, 10.0, and 10.0, then in this case, the student would have an arithmetic average (mean) of 8.0, a median score of 9.0, and a mode of 10.0.

If a student had six pop quizzes during the semester and received scores of 4.0, 4.0, 6.0, 8.0, 10.0, and 10.0, then what would be the median and mode? The arithmetic average (mean) would be 7.0, the median also be 7.0 (average of the middle terms 6.0 and 8.0) and we would have two modes, 4.0 and 10.0. Note that the median is always an average of the two middle terms when we have an even set of data and we never average two or more different modes.

Range and standard deviation are measures of dispersion. Range is the difference between the smallest and largest values in a set of data. Standard deviation is a measure of how much variance there is among values in a set of data. In thinking about standard deviation in a broad and conceptual way, we only need remember that the greater the standard deviation, the more the values are dispersed from the mean. A high standard deviation tells us that all or most of the values are spread out; a low standard deviation tells us that all or most of the values are bunched together.

It is true to say that standard deviation is greatest for a heterogeneous group of items and lowest for a homogeneous group of items. A homogeneous group of data is similar in the characteristic we are observing. A heterogeneous group of data is dissimilar in the characteristic we are observing. Say the characteristic we are observing is height. We would expect the standard deviation for the height of a team of professional basketball players to be lower than it would be for a group of people randomly selected at the local mall. Among a team of basketball players, tallness is nothing unusual. In visualizing the standard deviation for a team of such players, think first in terms of the estimated mean (arithmetic) for the height of the players. Once we visualize the mean height, we can see that most players approximate this height, although there certainly are a few shorter players and a few really tall players.

Now suppose we randomly survey the heights of people we find at the local mall. Our sample consists of children, teenagers, and adults across both genders. Obviously, we can expect a far greater standard deviation among individuals selected at random at the mall than the players playing on a professional basketball team. It is equally true to say that the heights of the team players are more homogeneous relative to the heights of people surveyed at the mall, and that the heights of people surveyed at the mall are more heterogeneous relative to the heights of a team of basketball players.

Exhibit 4.3   Consumer Age and Purchasing Patterns

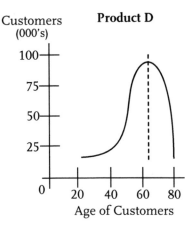

The purchase/sales patterns of four different products in a department store were tabulated over a fixed period of time, according to customers' ages, ranging between 20 and 80 years, and the results are displayed in the graphs in exhibit 4.3. Answer the following six questions with respect to the customers' ages.

94. Trends Q1 (🌶)

The graph of which individual product(s) shows an identical mean, median, and mode?

95. Trends Q2 (🌶)

The graph of which of the four products shows the smallest mode?

96. Trends Q3 (🌶 🌶)

The graph of which product shows a greater mean compared with its mode?

97. Trends Q4 (🌶 🌶)

The graph of which product shows a greater mode compared with its mean?

98. Trends Q5 (🌶)

The graph of which product shows the smallest range?

99. Trends Q6 (🌶)

The graph of which product shows the lowest standard deviation?

Statistical Significance

🖝 *Tip #30:* *Just because two values are different does not mean that they are statistically significantly different. And just because two values are statistically significantly different does not mean that their difference is substantively meaningful.*

Snapshot

A very interesting concept in statistics is the topic of statistical significance: that is, the point at which a difference really matters.

Just because two values are different does not mean that they are statistically significantly different.

You are surprised to read that a new study claims that vegetarians live on average up to 18 months longer than do non-vegetarians. A series of questions flood to mind. "Is this difference a big difference? Would I live longer if I became a vegetarian? Isn't living longer also due to chance regardless of one's eating habits?

Statistics (the science of analyzing data and drawing conclusions, often for the purpose of making predictions) comes to the rescue as if to say, "Give me your sample and let's do some calculations, and then we'll be able to tell what this 18-month difference really means." At what point, measured in months or years, does a difference in longevity really matter?

If the difference of 18 months is statistically significant, it would mean that this difference is not due merely to chance but that being a vegetarian really does increase one's longevity. If an 18-month difference is not statistically significant, it would mean that this difference is not enough to link vegetarianism to increased longevity. It could be that chance itself is all that is needed to explain why vegetarians in the study do or do not live a year and a half longer than does the average non-vegetarian.

The phrase "statistical significance" has a special meaning in statistics. It is linked to the figure of 95%. If a difference is statistically significant, it is 95% probable that this difference is genuine and not simply a matter of chance. The 95% level is somewhat mysterious but likely comes from academia, where a theory usually has to have at least a 95% chance of being true to be considered worth telling people about. According to our vegetarian study, if the difference is statistically significant, we could believe with 95% certainty that a switch to vegetarianism would help us increase our longevity on average by one and a half years.

A few additional points. First, in normal English, "significant" means important. In statistics, "significant" means probably true (that is, not due to chance). Statistically significant therefore does not necessarily mean important. When a statistic is significant, it doesn't necessarily mean that the finding is important or interesting, or that it has any decision-making utility.

Second, although we have cited the 95% figure, most often, a 5% figure is used in discussing statistical significance. The good news is that these numbers are just mirror opposites of each other. A finding is statistically significant if there is less than a 5% chance that it's not true (or less than a 5% chance that the difference is due to chance). In short, it doesn't matter whether we say that there is a 95% or better chance that something is true or that there is less than a 5% chance that something is not true.

Third, we are proceeding under the assumption that sample sizes derived from studies or surveys are sufficiently large. If a sample size is too small, any claim of a real difference between two values will be inconclusive.

Just because two values are statistically significantly different does not mean that their difference is substantively meaningful.

Even if an 18-month increase in longevity is a statistically significant difference, a person might not conclude that this difference in lifespan is substantively meaningful. That is, a person might not be moved to become a vegetarian because he or she finds the practice too inconvenient and restrictive (e.g., difficulty of finding vegetarian food and/or hassle when ordering in the company of non-vegetarians).

Sometimes the prospect of adverse consequences must be factored in as a qualitative component of our decision-making process. Say a pharmaceutical company develops a drug to help with cancer, and an outside agency surveys the lifespan of people who do and do not take the drug and concludes that statistics show that the drug enhances lifespan by four weeks. Statistically speaking, that augmented lifespan may be significantly longer for patients who took the drug, but is the difference substantively meaningful? Does that extra four weeks merit the drug's cost and potential side effects? Although statistics may tell us that a given difference matters, it still might not be a meaningful difference, particularly in light of other considerations.

100. Statistician (🌶🌶)

If the difference between two numbers or values is deemed statistically significant, then which of the following must be true?

I.   The difference between these two numbers or values is, by implication, considered important.

II.   The difference between these two numbers or values cannot at the same time also be considered not statistically significant.

III. The odds that the difference between these two numbers or values cannot be relied upon are less than a 1 in 20 chance.

IV.   There is at least a 95% chance that the difference between these two numbers or values can be relied upon, and less than a 5% chance that this difference is due to chance.

A)   I & II
B)   III & IV
C)   I, II & III
D)   II, III & IV
E)   All of the above

Appendixes

The beauty of math is that things always add up. Success in math is 70 percent hard work, 30 percent skill, and 10 percent luck.

—Anonymous

Appendix I – Numeracy Principles 1 to 30

Percentages

🖝 Tip #1: A percentage increase is not the same as a percentage of an original number.

🖝 Tip #2: You can't add (or subtract) the percentages of different wholes.

🖝 Tip #3: Percentages cannot be compared directly to numbers because percentages are relative measures while numbers are actual values.

🖝 Tip #4: Percentages (or ratios) are necessary tools for drawing conclusions about qualitative measures, such as preferability, danger, or safety.

🖝 Tip #5: Growing pies presents a classic case where the dollar value of a given item increases over time, yet this same item decreases in percentage terms relative to a larger pie.

🖝 Tip #6: Consider the following statement: "Person A spends a greater portion of his or her disposable income on beer than on coffee as compared to person B." This does not necessarily mean that person A spends more money in total on beer and coffee than does person B. It also does not mean that Person A, individually, spends more money on beer than he or she does on coffee.

Ratios & Proportions

🖝 Tip #7: Ratios tell us nothing about actual size or value.

☞ Tip #8:   You cannot add or subtract a number from a ratio unless you know the exact numbers that make up the ratio. However, you can always multiply or divide both sides of a ratio by a given number.

☞ Tip #9:   A part-to-part ratio is not the same thing as a part-to-whole ratio. If the ratio of female to male attendees at a conference is 1:2, this does not mean that 50 percent of the conference attendees are female.

☞ Tip #10:   Proportions, in the form of $\frac{A}{B} = \frac{C}{D}$, are among the simplest but most versatile math problem-solving tools.

☞ Tip #11:   If product A is selling for 20% more than product B, then the ratio of product A to product B is 120% to 100% (not 100% to 80%). If product B is selling for 20% less than product A, then the ratio of product A to product B is 100% to 80% (not 120% to 100%).

Reciprocals

☞ Tip #12:   Dividing by any number produces the same result as multiplying by that number's reciprocal (and vice-versa).

Order of Operations

☞ Tip #13:   The order in which two or more numbers are added or multiplied does not matter. However, but the order in which two or more numbers are subtracted or divided does matter.

CHAPTER 2 – WONDERFUL MATH RECIPES

Overlap Scenarios

🖙 Tip #14: Situations involving either A or B do not necessarily preclude the possibility of both A and B. When two events overlap, the number of items in both groups will exceed 100% due to mutual inclusivity.

Matrix Scenarios

🖙 Tip #15: A matrix is a marvelous tool for summarizing data that is being contrasted across two variables and that can be sorted into four distinct outcomes.

Mixture Scenarios

🖙 Tip #16: Mixture problems are best solved using the barrel method, which enables information to be summarized in a 3-row by 3-column table.

Weighted Average Scenarios

🖙 Tip #17: To find a weighted average, multiply each event by its associated weight and total the results. In the case of probabilities, multiply each event by its respective probability and total the results.

Chapter 3 – Favorite Numeracy Dishes

Markup vs. Margin

☞ Tip #18:   Markup is always larger than margin.

Price, Cost, Volume, and Profit

☞ Tip #19:   An increase (decrease) in sales volume does not necessarily translate to an increase (decrease) in profit.

☞ Tip #20:   When price-cost-volume-profit problems are presented in the form of algebraic expressions, it is best to first find a dollar per unit, dollar per person, usage per unit, or usage per person figure.

Break-Even Point

☞ Tip #21:   In break-even point problems, break-even occurs exactly where variable revenue (sales revenue less variable costs) equals total fixed costs.

Aggregate Costs vs. Per-Unit Costs

☞ Tip #22:   Total costs must be distinguished from per-unit costs, as is the case whenever the number of units varies between two situations under comparison.

Efficiency

☞ Tip #23:   Efficiency is output divided by input. Cost efficiency is cost divided by efficiency.

Distribution and Allocation

☞ Tip #24: Elements within distributions may be uniformly spaced or concentrated in "bunches." Allocations may be equal or unequal, proportionate or disproportionate.

☞ Tip #25: Front-loading occurs when a greater proportion of inputs (i.e., time, materials, costs, efforts, benefits) are entered early in a process. Back-loading occurs when a greater proportion of inputs are entered late in a process.

CHAPTER 4 – SPECIAL MATH GARNISHMENTS

Translating Simple Graphs

☞ Tip #26: Forward-slashing graphs indicate that two variables are directly related. Back-slashing graphs indicate that two variables are indirectly (inversely) related. Graphs formed with straight lines have variables which are linearly related. Graphs formed with curved lines have variables which are non-linearly related.

Line Graphs, Pie Charts, and Bar Charts

☞ Tip #27: Line graphs are best used to show how data changes over time. Pie charts are great for breaking data down by category. Bar charts are ideal for side-by-side comparisons of up to six pieces of data. Knowing the strength of each type of graph also helps us interpret the underlying data.

Correlation

➣ Tip #28:   Positive correlations have graphs sloping upward; negative correlations have graphs sloping downward. The stronger the correlation between two variables, the nearer to a straight line will be their graph; the graph of a perfect correlation is a straight line.

Measures of Average and Dispersion

➣ Tip #29:   The mean, median, and mode for a given set of data may be all different values. Data with a high standard deviation is more spread out, and the graph of such data is typically "low" and "wide." Data with a low standard deviation is bunched, and its graph is typically "narrow" and "tall."

Statistical Significance

➣ Tip #30:   Just because two values are different does not mean that they are statistically significantly different. And just because two values are statistically significantly different does not mean that their difference is substantively meaningful.

Appendix II – The World of Numbers

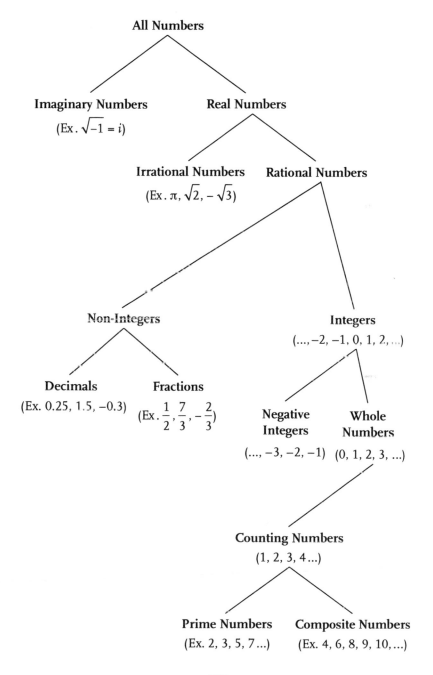

All Numbers

Imaginary Numbers
(Ex. $\sqrt{-1} = i$)

Real Numbers

Irrational Numbers
(Ex. π, $\sqrt{2}$, $-\sqrt{3}$)

Rational Numbers

Non-Integers

Integers
(..., -2, -1, 0, 1, 2, ...)

Decimals
(Ex. 0.25, 1.5, -0.3)

Fractions
(Ex. $\frac{1}{2}$, $\frac{7}{3}$, $-\frac{2}{3}$)

Negative
Integers
(..., -3, -2, -1)

Whole
Numbers
(0, 1, 2, 3, ...)

Counting Numbers
(1, 2, 3, 4...)

Prime Numbers
(Ex. 2, 3, 5, 7...)

Composite Numbers
(Ex. 4, 6, 8, 9, 10,...)

Numbers are first divided into real and imaginary numbers. Imaginary numbers are not a part of our everyday life. *Real* numbers are further divided into rational and irrational numbers. *Irrational* numbers are those which cannot be expressed as simple integers, fractions, or decimals; non-repeating decimals are always irrational numbers, of which π may be the most famous. *Rational* numbers include integers and non-integers. *Integers* are positive and negative whole numbers, while *non-integers* include decimals and fractions.

Number Definitions

Real numbers: Any number which exists on the number line. Real numbers are the combined group of rational and irrational numbers.

Imaginary numbers: Any number multiplied by i, the imaginary unit. $i = \sqrt{-1}$. Imaginary numbers are the opposite of real numbers and are not part of our everyday life.

Rational numbers: Numbers which can be expressed as a fraction whose top (the numerator) and bottom (the denominator) are both integers.

Irrational numbers: Numbers which can't be expressed as a fraction whose top (the numerator) and bottom (the denominator) are integers. Square roots of non-perfect squares (such as $\sqrt{2}$) are irrational, and π is irrational. Irrational numbers may be described as non-repeating decimals.

Integers: Integers consist of those numbers which are multiples of 1: $\{\ldots, -2, -1, 0, 1, 2, 3, \ldots\}$. Integers are "integral": they contain no fractional or decimal parts.

Non-integers: Non-integers are those numbers which contain fractional or decimal parts. For example, $\frac{1}{2}$ and 0.125 are non-integers.

Whole numbers: Non-negative integers: $\{0, 1, 2, 3, ...\}$. Note that 0 is a whole number (but a non-negative whole number, since it is neither positive nor negative).

Counting numbers: The subset of whole numbers which excludes 0: $\{1, 2, 3, ...\}$.

Prime numbers: Prime numbers are a subset of the counting numbers. They include those non-negative integers which have two and only two factors; that is, the factors 1 and themselves. The first 10 primes are 2, 3, 5, 7, 11, 13, 17, 19, 23, and 29. Note that 1 is not a prime number as it only has one factor, i.e., 1. Also, the number 2 is not only the smallest prime but also the only even prime number.

Composite numbers: A positive number that has more than two factors other than 1 and itself. Also, any non-prime number greater than 1. Examples include: 4, 6, 8, 9, and 10. Note that 1 is not a composite number, and the number 4 is the smallest composite number.

Factors: A factor is an integer that can be divided evenly into another integer ("divided evenly" means that there is no remainder). For example, the factors of 12 are 1, 2, 3, 4, 6, and 12.

Multiples: A multiple is a number that results from a given integer's being multiplied by another integer. Example: Multiples of 12 include 12, 24, 36, and 48. Proof: $12 \times 1 = 12$, $12 \times 2 = 24$, $12 \times 3 = 36$, and $12 \times 4 = 48$. Note that whereas a factor of any number is less than or equal to the number in question, a multiple of any number is equal to or greater than the number itself. That is, any non-zero integer has a finite number of factors but an infinite number of multiples.

The Four Basic Operations

The four basic arithmetic operations are addition, subtraction, multiplication, and division. The results of these operations are called sum, difference, product, and quotient, respectively. Two additional operations involve exponents and radicals.

Addition

Subtraction

Multiplication

Division

Exponents

Radicals

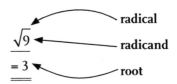

Common Fractions and their Percentage Equivalents

Exercise – Fill in the missing percentages to change common fractions into their percentage equivalents.

$$\frac{1}{1} = 100\%$$

$$\frac{1}{2} = 50\% \qquad \frac{2}{2} = 100\%$$

$$\frac{1}{3} = ?\% \qquad \frac{2}{3} = ?\% \qquad \frac{3}{3} = 100\%$$

$$\frac{1}{4} = ?\% \qquad \frac{2}{4} = 50\% \qquad \frac{3}{4} = ?\% \qquad \frac{4}{4} = 100\%$$

$$\frac{1}{5} = ?\% \qquad \frac{2}{5} = ?\% \qquad \frac{3}{5} = ?\% \qquad \frac{4}{5} = ?\% \qquad \frac{5}{5} = 100\%$$

$$\frac{1}{6} = ?\% \qquad \frac{2}{6} = ?\% \qquad \frac{3}{6} = 50\% \qquad \frac{4}{6} = ?\% \qquad \frac{5}{6} = ?\% \qquad \frac{6}{6} = 100\%$$

$$\frac{1}{7} = ?\% \qquad \frac{2}{7} = ?\% \qquad \frac{3}{7} = ?\% \qquad \frac{4}{7} = ?\% \qquad \frac{5}{7} = ?\% \qquad \frac{6}{7} = ?\% \qquad \frac{7}{7} = 100\%$$

$$\frac{1}{8} = ?\% \qquad \frac{2}{8} = ?\% \qquad \frac{3}{8} = ?\% \qquad \frac{4}{8} = 50\% \qquad \frac{5}{8} = ?\% \qquad \frac{6}{8} = ?\% \qquad \frac{7}{8} = ?\% \qquad \frac{8}{8} = 100\%$$

$$\frac{1}{9} = ?\% \qquad \frac{2}{9} = ?\% \qquad \frac{3}{9} = ?\% \qquad \frac{4}{9} = ?\% \qquad \frac{5}{9} = ?\% \qquad \frac{6}{9} = ?\% \qquad \frac{7}{9} = ?\% \qquad \frac{8}{9} = ?\% \qquad \frac{9}{9} = 100\%$$

$$\frac{1}{10} = ?\% \qquad \frac{2}{10} = ? \qquad \frac{3}{10} = ?\% \qquad \frac{4}{10} = ? \qquad \frac{5}{10} = 50\% \qquad \frac{6}{10} = ? \qquad \frac{7}{10} = ? \qquad \frac{8}{10} = ? \qquad \frac{9}{10} = ? \qquad \frac{10}{10} = 100\%$$

Solutions – Common fractions and their percentage equivalents.

$\frac{1}{1} = 100\%$

$\frac{1}{2} = 50\%$ $\frac{2}{2} = 100\%$

$\frac{1}{3} = 33\frac{1}{3}\%$ $\frac{2}{3} = 66\frac{2}{3}\%$ $\frac{3}{3} = 100\%$

$\frac{1}{4} = 25\%$ $\frac{2}{4} = 50\%$ $\frac{3}{4} = 75\%$ $\frac{4}{4} = 100\%$

$\frac{1}{5} = 20\%$ $\frac{2}{5} = 40\%$ $\frac{3}{5} = 60\%$ $\frac{4}{5} = 80\%$ $\frac{5}{5} = 100\%$

$\frac{1}{6} = 16\frac{2}{3}\%$ $\frac{2}{6} = 33\frac{1}{3}\%$ $\frac{3}{6} = 50\%$ $\frac{4}{6} = 66\frac{2}{3}\%$ $\frac{5}{6} = 83\frac{1}{3}\%$ $\frac{6}{6} = 100\%$

$\frac{1}{7} = 14.3\%$ $\frac{2}{7} = 28.6\%$ $\frac{3}{7} = 42.9\%$ $\frac{4}{7} = 57.1\%$ $\frac{5}{7} = 71.4\%$ $\frac{6}{7} = 85.7\%$ $\frac{7}{7} = 100\%$

$\frac{1}{8} = 12.5\%$ $\frac{2}{8} = 25\%$ $\frac{3}{8} = 37.5\%$ $\frac{4}{8} = 50\%$ $\frac{5}{8} = 62.5\%$ $\frac{6}{8} = 75\%$ $\frac{7}{8} = 87.5\%$ $\frac{8}{8} = 100\%$

$\frac{1}{9} = 11.11\%$ $\frac{2}{9} = 22.22\%$ $\frac{3}{9} = 33.33\%$ $\frac{4}{9} = 44.44\%$ $\frac{5}{9} = 55.55\%$ $\frac{6}{9} = 66.66\%$ $\frac{7}{9} = 77.77\%$ $\frac{8}{9} = 88.88\%$ $\frac{9}{9} = 100\%$

$\frac{1}{10} = 10\%$ $\frac{2}{10} = 20\%$ $\frac{3}{10} = 30\%$ $\frac{4}{10} = 40\%$ $\frac{5}{10} = 50\%$ $\frac{6}{10} = 60\%$ $\frac{7}{10} = 70\%$ $\frac{8}{10} = 80\%$ $\frac{9}{10} = 90\%$ $\frac{10}{10} = 100\%$

Rules for Odd and Even Numbers

Even + Even = Even

    Example: $2 + 2 = 4$
    Example: $-2 + -2 = -4$

Odd + Odd = Even

    Example: $3 + 3 = 6$
    Example: $-3 + -3 = -6$

Even + Odd = Odd

    Example: $2 + 3 = 5$
    Example: $-2 + -3 = -5$

Odd + Even = Odd

    Example: $3 + 2 = 5$
    Example: $-3 + -2 = -5$

Even − Even = Even

    Example: $4 - 2 = 2$
    Example: $-4 - (-2) = -2$

Odd − Odd = Even

    Example: $5 - 3 = 2$
    Example: $-5 - (-3) = -2$

Even − Odd = Odd

    Example: $6 - 3 = 3$
    Example: $-6 - (-3) = -3$

Odd − Even = Odd

    Example: $5 - 2 = 3$
    Example: $-5 - (-2) = -3$

Even × Even = Even

    Example: $2 \times 2 = 4$
    Example: $-2 \times -2 = 4$

Odd × Odd = Odd

    Example: $3 \times 3 = 9$
    Example: $-3 \times -3 = 9$

Even × Odd = Even

    Example: $2 \times 3 = 6$
    Example: $-2 \times -3 = 6$

Odd × Even = Even

    Example: $3 \times 2 = 6$
    Example: $-3 \times -2 = 6$

Even ÷ Even = Even

    Example: $4 \div 2 = 2$
    Example: $-4 \div -2 = 2$

Odd ÷ Odd = Odd

    Example: $9 \div 3 = 3$
    Example: $-9 \div -3 = 3$

Even ÷ Odd = Even

    Example: $6 \div 3 = 2$
    Example: $-6 \div -3 = 2$

Odd ÷ Even = *Not Possible

Example: $5 \div 2 = 2\frac{1}{2}$
Example: $-5 \div -2 = 2\frac{1}{2}$

* Per the previous two examples, an odd number divided by an even number does not result in either an even or odd integer; it results in a non-integer.

Rules for Positive and Negative Numbers

Positive + Positive = Positive

Example: $2 + 2 = 4$

Negative + Negative = Negative

Example: $-2 + (-2) = -4$

Positive + Negative = Depends

Example: $4 + (-2) = 2$
Example: $2 + (-4) = -2$

Negative + Positive = Depends

Example: $-2 + 4 = 2$
Example: $-4 + 2 = -2$

Positive − Positive = Depends

Example: $4 - 2 = 2$
Example: $2 - 4 = -2$

Negative − Negative = Depends

Example: $-2 - (-4) = 2$
Example: $-4 - (-2) = -2$

Positive – Negative = Positive

Example: $2 - (-2) = 4$

Negative – Positive = Negative

Example: $-4 - 2 = -6$

Positive × Positive = Positive

Example: $2 \times 2 = 4$

Negative × Negative = Positive

Example: $-2 \times -2 = 4$

Positive × Negative = Negative

Example: $2 \times -2 = -4$

Negative × Positive = Negative

Example: $-2 \times 2 = -4$

Positive ÷ Positive = Positive

Example: $4 \div 2 = 2$

Negative ÷ Negative = Positive

Example: $-4 \div -2 = 2$

Positive ÷ Negative = Negative

Example: $4 \div -2 = -2$

Negative ÷ Positive = Negative

Example: $-4 \div 2 = -2$

Common Squares, Cubes, and Square Roots

$1^2 = 1$	$1^3 = 1$	$1^4 = 1$	$1^5 = 1$	$1^6 = 1$
$2^2 = 4$	$2^3 = 8$	$2^4 = 16$	$2^5 = 32$	$2^6 = 64$
$3^2 = 9$	$3^3 = 27$	$3^4 = 81$	$3^5 = 243$	
$4^2 = 16$	$4^3 = 64$	$4^4 = 256$		
$5^2 = 25$	$5^3 = 125$	$5^4 = 625$		
$6^2 = 36$	$6^3 = 216$			
$7^2 = 49$	$7^3 = 343$			
$8^2 = 64$	$8^3 = 512$			
$9^2 = 81$	$9^3 = 729$			

Also:

$10^2 = 100$
$10^3 = 1,000$
$10^4 = 10,000$
$10^5 = 100,000$
$10^6 = 1,000,000$
$10^9 = 1$ billion
$10^{12} = 1$ trillion

Author's note: In most English-speaking countries today (particularly the U.S., Great Britain, Canada, and Australia), one billion equals 1,000,000,000 or 10^9, or one thousand millions. In many other countries, including France, Germany, Spain, Norway, and Sweden, the word "billion" indicates 10^{12}, or one million millions. Although Britain and Australia have traditionally employed the international usage of 10^{12}, they have now largely switched to the U.S. version of 10^9.

Common Squares from 13 to 30

$13^2 = 169$	$19^2 = 361$	$25^2 = 625$
$14^2 = 196$	$20^2 = 400$	$26^2 = 676$
$15^2 = 225$	$21^2 = 441$	$27^2 = 729$
$16^2 = 256$	$22^2 = 484$	$28^2 = 784$
$17^2 = 289$	$23^2 = 529$	$29^2 = 841$
$18^2 = 324$	$24^2 = 576$	$30^2 = 900$

Common Square Roots

$$\sqrt{1} = 1$$
$$\sqrt{2} = 1.4$$
$$\sqrt{3} = 1.7$$
$$\sqrt{4} = 2$$
$$\sqrt{5} = 2.2$$

Pop Quiz

See pages 153–154 for answers.

Review – Fractions to Percents

Convert the following fractions to their percentage equivalents:

$\frac{1}{3} =$	$\frac{2}{3} =$	$\frac{1}{6} =$	$\frac{5}{6} =$	$\frac{1}{8} =$
$\frac{3}{8} =$	$\frac{5}{8} =$	$\frac{7}{8} =$	$\frac{1}{9} =$	$\frac{5}{9} =$

Review – Decimals to Fractions

Convert the following decimals, which are greater than 1, into fractional equivalents.

1.2 =   1.25 =   1.33 =

Simplify each expression below without multiplying decimals or using a calculator. (Hint: Translate each decimal into a common fraction and then multiply and/or divide.)

(1.25)(0.50)(0.80)(2.00) =

$$\frac{(0.7500)(0.8333)}{(0.6250)} =$$

$$\frac{(0.2222)}{(0.3333)(0.6666)} =$$

Review – Common Square Roots

Put the following statements in order, from largest to smallest value. (Hint: Approximate each square root to one decimal point.)

I. $1 + \sqrt{5}$
II. $2 + \sqrt{3}$
III. $3 + \sqrt{2}$

Divisibility Rules

No.	Divisibility Rule	Examples
1	Every number is divisible by 1.	15 divided by 1 equals 15.
2	A number is divisible by 2 if it is even.	24 divided by 2 equals 12.
3	A number is divisible by 3 if the sum of its digits is divisible by 3.	651 is divisible by 3 since 6 + 5 + 1 = 12 and "12" is divisible by 3.
4	A number is divisible by 4 if its last two digits form a number that is divisible by 4.	1,112 is divisible by 4 since the number "12" is divisible by 4.
5	A number is divisible by 5 if the number ends in 5 or 0.	245 is divisible by 5 since this number ends in 5.
6	A number is divisible by 6 if it is divisible by both 2 and 3.	738 is divisible by 6 since this number is divisible by both 2 and 3, and the rules that govern the divisibility of 2 and 3 apply.
7	No clear rule.	N.A.
8	A number is divisible by 8 if its last three digits form a number that is divisible by 8.	2,104 is divisible by 8 since the number "104" is divisible by 8.
9	A number is divisible by 9 if the sum of its digits is divisible by 9.	4,887 is divisible by 9 since 4 + 8 + 8 + 7 = 27 and 27 is divisible by 9.
10	A number is divisible by 10 if it ends in 0.	990 is divisible by 10 because 990 ends in 0.

Answers and Explanations

Review – Fractions to Percents

$$\frac{1}{3} = 33\frac{1}{3}\% \qquad \frac{2}{3} = 66\frac{2}{3}\% \qquad \frac{1}{6} = 16\frac{2}{3}\% \qquad \frac{5}{6} = 83\frac{1}{3}\%$$

$$\frac{1}{8} = 12.5\% \qquad \frac{3}{8} = 37.5\% \qquad \frac{5}{8} = 62.5\% \qquad \frac{7}{8} = 87.5\%$$

$$\frac{1}{9} = 11.11\% \qquad \frac{5}{9} = 55.55\%$$

Review – Decimals to Fractions

The answer to the first question is found by adding 1 to the fractional equivalent of 0.2.

$$0.2 = \frac{1}{5} \quad \text{Thus, } 1.2 = 1 + \frac{1}{5} = 1\frac{1}{5} = \frac{6}{5}$$

The answer to the second question is found by adding 1 to the fractional equivalent of 0.25.

$$0.25 = \frac{1}{4} \quad \text{Thus, } 1.25 = 1 + \frac{1}{4} = 1\frac{1}{4} = \frac{5}{4}$$

The answer to the third question is found by adding 1 to the fractional equivalent of 0.33.

$$0.33 = \frac{1}{3} \quad \text{Thus, } 1.33 = 1 + \frac{1}{3} = 1\frac{1}{3} = \frac{4}{3}$$

Each of the following three "fraction" problems equals 1!

$$\frac{5}{4} \times \frac{5}{10} \times \frac{8}{10} \times \frac{2}{1} = \frac{400}{400} = 1$$

$$\frac{\dfrac{3}{4} \times \dfrac{5}{6}}{\dfrac{5}{8}} = \frac{\dfrac{15}{24}}{\dfrac{5}{8}} = \frac{15}{24} \times \frac{8}{5} = \frac{120}{120} = 1$$

$$\frac{\dfrac{2}{9}}{\dfrac{1}{3} \times \dfrac{2}{3}} = \frac{\dfrac{2}{9}}{\dfrac{2}{9}} = \frac{2}{9} \times \frac{9}{2} = \frac{18}{18} = 1$$

Review – Common Square Roots

The answer to this square root problem is III, II & I. These statements are, in fact, in order from smallest to largest value, but the question requires ordering the values from largest to smallest.

I. $1 + 2.2 = 3.2$
II. $2 + 1.7 = 3.7$
III. $3 + 1.4 = 4.4$

CHAPTER 1 – BASIC NUMERACY INGREDIENTS

1. Antique Stamp ( )

Choice E. This is a "percentage of an original number" problem. This problem is essentially asking how many times larger, in percentage terms, is the current price compared with the original price.

$$\frac{\text{New}}{\text{Old}} \qquad \frac{\$150}{\$100} = 150\%$$

Author's note: The calculation below is technically more accurate. The dollar signs (i.e., $) cancel out and 1.5 must be multiplied by 100% in order to turn this decimal into a percentage and to reinstate the percentage sign in the final answer.

$$\frac{\$150}{\$100} = 1.5 \times 100\% = 150\%$$

2. Percentage Increase (🌶)

Choice D. This is a straight percentage increase. The formula for a percentage increase is as follows:

$$\frac{\text{New} - \text{Old}}{\text{Old}} \qquad \frac{\$150 - \$100}{\$100} = \frac{\$50}{\$100} = 50\%$$

3. Percentage Decrease (🌶)

Choice D. This is a straight percentage decrease. The formula for a percentage decrease is as follows:

$$\frac{\text{Old} - \text{New}}{\text{Old}} \qquad \frac{\$100 - \$50}{\$100} = \frac{\$50}{\$100} = 50\%$$

4. Devalue (🌶)

Choice A. The formula for a percentage decrease to return to an original number is as follows:

$$\frac{\text{New} - \text{Old}}{\text{New}} \qquad \frac{\$125 - \$100}{\$125} = \frac{\$25}{\$125} = 20\%$$

5. Revalue ()

Choice C. The formula for a percentage increase to return to an original number is as follows:

$$\frac{\text{Old} - \text{New}}{\text{New}} \qquad \frac{\$100 - \$75}{\$75} = \frac{\$25}{\$75} = 33\frac{1}{3}\%$$

6. Microbrewery (𝕊𝕊)

Choice C. This problem is in essence asking about productivity: productivity = output ÷ hours. Use 100% as a base, and add 70% to get 170%; then divide 170% by 80% to get 212.5%. An even simpler calculation involves the use of decimals:

$$\frac{1.7}{0.8} = 2.125 \times 100\% = 212.5\%$$

Now calculate the percentage increase:

$$\frac{\text{New} - \text{Old}}{\text{Old}} \qquad \frac{212.5\% - 100\%}{100\%} = \frac{112.5\%}{100\%} = 112.5\%$$

Encore! What if the wording to this problem had been identical except that the last sentence read:

The year-end factory output per hour is what percentage of the factory output per hour at the beginning of the year?

A)   50%
B)   90%
C)   112.5%
D)   210%
E)   212.5%

The answer would be choice E. This problem is not asking for a percentage increase, but rather "percentage of an original number."

Percentage of an original number:

$$\frac{\text{New}}{\text{Old}} \quad \frac{212.5\%}{100\%} = 212.5\%$$

Again, the calculation below is more accurate. The percent signs (i.e., %) cancel out and 2.125 must be multiplied by 100% in order to turn this decimal into a percentage and to reinstate the percentage sign in the final answer.

$$\frac{212.5\%}{100\%} = 2.125 \times 100\% = 212.5\%$$

Author's note: The fact that two of the answer choices, namely choices C and E, are 100% apart alerts us to the likelihood that a distinction needs to be made between "percentage increase" and "percentage of an original number."

7. Gardener (🌿 🌿)

Choice B. View the area of the original rectangular garden as having each a width and length of 100%. The new rectangular garden has a length of 140% and a width of 80%. A 20% decrease in width translates to a width of 80% of the original.

Area of original garden:

Area = length × width = (100% × 100%) = 100%

Area of new garden:

Area = length × width = (140% × 80%) = 112%

Percentage change is as follows:

$$\frac{\text{New} - \text{Old}}{\text{Old}} = \frac{(112\% - 100\%)}{100\%} = \frac{12\%}{100\%} = 12\%$$

Note: This shortcut calculation involves decimals:

$1.4 \times 0.8 = 2.12$
$2.12 - 1.0 = 1.12 \times 100\% = 112\%$

8. Train ( )

Choice C. In this problem, we must stick to the percentage increase formula. The correct calculation for the new speed is:

$125\% \times 120\% = 150\%$

$$\frac{\text{New} - \text{Old}}{\text{Old}} \qquad \frac{150\% - 100\%}{100\%} = \frac{50\%}{100\%} = 50\%$$

There are two trap answers. In choice B, we cannot just add 25% and 20% to get 45% because 20% of 125% is a bigger number than 20% of 100%. In choice E, 150% represents percentage of an original number, not percentage increase.

9. Double Discount ( )

Choice D.

$80\% \times 90\% = 72\%$
$100\% - 72\% = 28\%$

A 20% discount followed by a 10% discount amounts to a 28% discount. We cannot simply add 20% to 10% to get 30%. The rationale for this is encapsulated by tip #2—You can't add (or subtract) the percentages of different wholes. The first whole is 100% and the second whole is 80%.

10. Net Effect #1 (🌶)

Choice C.

> 90% × 110% = 99%
> 100% − 99% = 1% net decrease

Say that the original price was 100%. It decreases from 100% to 90%. When it increases by 10%, it is now 99% of the original price. This works out simply because 10% of 100% is a greater percentage than 10% of 90%. Note that using decimals rather than percentages makes things even simpler.

> $0.9 \times 1.1 = 0.99$
> $1.0 - 0.99 = 0.01 \times 100\% = 1\%$ net decrease

11. Net Effect #2 (🌶)

Choice C.

> 120% × 80% = 96%
> 100% − 96% = 4% net decrease

Say that the original price was 100%. It increases from 100% to 120%. When it decreases by 20%, it is now 96% of the original price. This works out simply because 20% of 120% is a greater percentage than 20% of 100%. Again, the use of decimals rather than percents is simpler.

> $1.2 \times 0.8 = 0.96$
> $1.0 - 0.96 = 0.04 \times 100\% = 4\%$ net decrease

12. Silver (🌶)

Choice C. Let's use 100% as a base. A 100% increase gives us 200%. A 50% decrease from 200% gives us 100%—the original base. The reason things work out this way is that the 100% increase is based on a smaller base (i.e., 100%) than is the 50% decrease, which is based on a larger base (i.e., 200%).

Quick calculation: $2.00 \times 0.5 = 100$

13. Rule of Finance (🌶)

Choice A. In fact, it will take approximately 7 years.

Calculation for $1@10% for 7 years:
$$\$1 \times (1.10)^7 = \$1 \times 1.95 \cong \$2$$

Calculation for $1@7% for 10 years:
$$\$1 \times (1.07)^{10} = \$1 \times 1.96 \cong \$2$$

The "Seven-Ten" Rule of Finance is so called because a business growing at 10% a year will double in size in 7 years. Obviously this is the result of compounded growth. Likewise, a business growing at 7% a year will double in size in approximately 10 years. The trap answer is 10 years—10 years at 10% per year equals 100%. But because the company is growing at 10% per year, the growth is compounding, and the 10% figure is being calculated on a different whole (i.e., principal plus interest) for each succeeding year.

14. Medical School (🌶 🌶)

Choice C. Percents and numbers do not always mix. Seventy-five percent cannot be compared with half of all female applicants. This question only makes sense if we know the percentage of females accepted and compare this figure with the percentage of males accepted. A likely scenario is that 75% of all students in medical school are male because 75% of all medical school applicants are male; 25% of all students accepted to medical school are female because 25% of the total applicants are female—little surprise.

Take this hypothetical example. Say that 10,000 applicants apply for a spot at a top medical school. What if 25% of the applicant pool is female while 75% is male? This means that 2,500 females and 7,500 males will apply for admission.

The opening class will consist of 500 students, and only 1 in 20 applicants, or 5% (i.e., 50/10,000), is accepted. Common sense tells us that with rules of fair selection in place, the opening class should be 75% male and 25% female, which directly reflects the applicant pool. Thus, the number of females in the opening class should be 125 (500 × 25%) while the number of males in the opening class should be 375 (500 × 75%).

Now, relate hypothetical figures within the context of the original passage. "The facts in the report speak for themselves: Seventy-five percent of all students in medical school are male, but fewer than half of all the 2,500 female applicants reach their goal of being admitted to medical school." Of course less than half of all females are admitted; only 1 in 20 students overall is accepted!

Choices A, B, D, and E are all irrelevant to the issue at hand. In choice B, there is no reason to suspect that the yield rates (percentage of offers given by medical schools that are accepted by students) would in any significant way be affected by gender. So although it might warrant some consideration, it is by no means a close contender with choice C.

15. White-Collar Crime (🎣 🎣)

Choice E. The number of white-collar crimes last year is not the same as the rate of increase in white-collar crime last year. Say, for example, that there were 900 cases of white-collar crime the year before last and there were 1,000 cases of white-collar crime this past year. This means that 50% of the increase, or 50 cases, is due to unemployment, often aggravated by alcoholism.

In terms of the number of white-collar crimes that occurred last year, most of them occurred due to desperation that results from overwork. This means that more than half of the 1,000 cases last year, or 500-plus cases, were due to loneliness that results from overwork.

The author's criticism of Dr. Noah is not valid because we cannot compare the 50 cases representing the percentage of rate increase with the 500 cases representing white-collar crime due to the desperation that results from overwork. In other words, the majority of white-collar crime is still due to the desperation that results from overwork, but the greatest increase (trend) has been caused by unemployment aggravated by alcoholism.

16. Fly Fishing (🎣 🎣)

Choice E. Percentages are really fractions. Fractions are composed of a numerator (top number in a faction) and denominator (bottom number in a fraction). A percentage is increased either by increasing the numerator of that fraction or by decreasing a denominator of that fraction, or doing both simultaneously. Choices A, B, C, and D could be true. However, choice E could not be true. Choice E must lead to a percentage decrease, not a percentage increase.

A) $\dfrac{\uparrow \text{ number of people fly-fishing}}{\downarrow \text{ sport-fishing population}} = \uparrow \%$

B) $$\frac{\uparrow \text{ number of people fly-fishing}}{\text{sport-fishing population (remains the same)}} = \uparrow \%$$

C) $$\frac{\uparrow \text{ number of people fly-fishing}}{\uparrow \text{ sport-fishing population}} = \uparrow \%$$

> Could be true as long as the numerator increases incrementally faster than the denominator.

D) $$\frac{\downarrow \text{ number of people fly-fishing}}{\downarrow \text{ sport-fishing population}} = \uparrow \%$$

> Could be true as long as the denominator decreases incrementally faster than the numerator.

E) $$\frac{\text{same number of people fly-fishing}}{\uparrow \text{ sport-fishing population}} = \downarrow \%$$

17. Lottery ( )

Choice C. We can't tell. It all depends on how many people are in each income category. For example, since we expect fewer people to be in the over $100,000 income category (compared with the other income categories), we also expect fewer ticket purchases by individuals in this category. This makes it impossible to know the ratio or percentage of people in each category who buy lottery tickets. In order to make an actual, meaningful comparison, we would need to divide the number of lottery tickets bought by individuals in each income category by the number of individuals in each category.

18. Iraq ( )

Choice D. Obviously the chances of an American solider being killed serving in Iraq during his or her first three-year period of the Second Gulf War were greater than the chances of a U.S. civilian driver being killed driving on the roads of America during the same three-year period. We cannot directly compare raw numbers. A more realistic comparison would be to compare combat deaths to the total number of combat soldiers while comparing traffic deaths to the total number of U.S. drivers. The following calculations show the role that each denominator plays in each emerging ratio:

$$\frac{\text{Military Combat Deaths}}{\text{Total Number of Active Soldiers in Iraq}} = \frac{2,300}{130,000} = 1.7\%$$

$$\frac{\text{Auto Traffic Deaths}}{\text{Total Number of US drivers}} = \frac{100,000}{50,000,000} = 0.2\%$$

From these statistics we see that it was 8.5 times more dangerous serving as a combatant in Iraq than it was to be a driver in America during the same three-year period.

$$\frac{\text{Death Rate (Military Combatants in Iraq)}}{\text{Death Rate (US Drivers)}} = \frac{1.7}{0.2} = 8.5$$

The scope of the two situations being compared is deaths serving in Iraq and deaths driving in America during a three-year period. The prospect of gun-related deaths (choice A) is essentially out of our argument's scope. And as deaths are unfortunately deaths, it is essentially irrelevant to consider "driving deaths resulting from high speeds" (choice B), "deaths caused by accidents" (choice C), or "percentage of time driving a car or serving in the armed forces" (choice E). This last choice is especially mind warping. Of course, it doesn't matter whether one drives an hour one day or serves a full day in Iraq. The argument is objectively comparing the death rate of one situation to the death rate of another. Besides, the fact that the average person drives for only a small portion of each day seems to help corroborate the original argument, not help refute it.

19. Military Expenditures (𝔖 𝔖)

Choice B. This problem highlights the concept of the growing pie. The apparent inconsistency is resolved if it is true that the portion or percentage of the national budget represented by military expenditures has increased by a greater amount than actual expenditures. If, for example, the national budget has increased tenfold during the past two decades and the amount of actual expenditures (inflation adjusted) has increased by 1 percent per year, then the portion devoted to military expenditure has likely shrunk. Even if actual dollars spent on military expenditures are increasing each year, as long as overall (or total) expenditures are increasing more, the amount of military expenditures, in percentage terms, is falling.

The opposite of a growing pie is a shrinking pie. If this problem were a shrinking pie, actual dollars spent on military expenditures would be decreasing each year, but as long as overall expenditures were shrinking faster, the amount of military expenditures, in percentage terms, would be rising.

20. Fiction Books (𝔖 𝔖 𝔖)

Choice E. This problem illustrates the ultimate tangling of numbers with percentages. Choice A is out of scope because we don't know about people with the highest levels of education; we only know about members of households with higher levels of education (i.e., HEL—high education level households) and members of households with lower levels of education (i.e., LEL—low education level households). Also, we do not know whether people really buy, are given, or inherit these books. Choices B, C, and D are out because we do not know for sure whether HEL households or LEL households have more or less fiction versus nonfiction books. We also do not know whether members of HEL households have more or fewer nonfiction books as compared with members of LEL households. The only thing we do know for sure is that if members of HEL households have more books in total than members of LEL households and a higher percentage of fiction versus nonfiction, then it must be true that HEL households contain a greater <u>number</u> of fiction books than LEL households.

All of the following are significant possibilities based on hypothetical numbers of books and percentages. Again, upon examination, the only thing that must be true is that for any given scenario the number of fiction books in HEL households must be greater than the number of fiction books in LEL households. Let's use 100 books and 50 books to test things out. The number of fiction books for HEL households, column (1), will always be greater than the number of fiction books for LEL households, column (3).

Scenario #1: HEL households have 80% fiction books and 20% nonfiction books while LEL households have 50% fiction books and 50% nonfiction books.

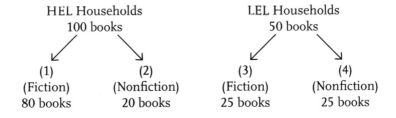

Scenario #2: HEL households have 70% fiction books and 30% nonfiction books while LEL households have 50% fiction books and 50% nonfiction books.

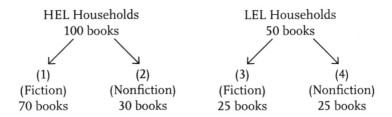

Scenario #3: Get ready for this! HEL households could have 40% fiction books and 60% nonfiction books while LEL households have 20% fiction books and 80% non-fiction books. This is possible because the percentage of fiction books for HEL households is still greater—40%—than the percentage of fiction books for LEL households—20%.

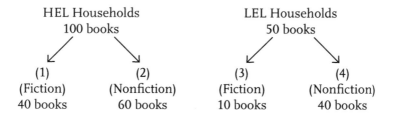

HEL Households 100 books		LEL Households 50 books	
(1) (Fiction) 40 books	(2) (Nonfiction) 60 books	(3) (Fiction) 10 books	(4) (Nonfiction) 40 books

21. WAA (𝄞 𝄞)

Choice A. We need the statement that would not provide a logical objection to the WAA's conclusion. Although it's probably true that countries with more miles of road are likely to have more traffic deaths, the ranking of deaths is per mile driven. This ratio makes it possible to compare various countries irrespective of how many miles of roadway they have.

The wrong answer choices are all valid objections because they point to reasons other than drivers' ability for the number of deaths per mile driven. The difference in road safety from one country to another (choice B) is a factor in accidents. How well a car is maintained (choice C) can also be a factor in whether a driver has an accident. A difference in terrain (choice D) means good drivers in mountainous countries may be more likely to have a fatal accident. And a higher number of passengers per car (choice E) might lead to a high number of traffic fatalities, even where the frequency of accidents is very low.

22. Scotch & Soda (🌶)

Choice E. The ratio cannot be determined. The trap answer is choice A. We cannot simply add the mixtures, as would be the case if the mixtures were of equal size (i.e., 1:2 + 2:3 = 3:5). If the original mixtures vary considerably in size, the final ratio will also vary considerably.

23. Eggs (🌶)

Choice D. As it is impossible to have 2.4 broken eggs (or 9.6 unbroken eggs), choice D is correct.

A) 1:1   12 eggs = 6 broken ($\frac{1}{2}$ × 12); 6 unbroken ($\frac{1}{2}$ × 12)

B) 1:2   12 eggs = 4 broken ($\frac{1}{3}$ × 12); 8 unbroken ($\frac{2}{3}$ × 12)

C) 1:3   12 eggs = 3 broken ($\frac{1}{4}$ × 12); 9 unbroken ($\frac{3}{4}$ × 12)

D) 1:4   12 eggs = 2.4 broken ($\frac{1}{5}$ × 12); 9.6 unbroken ($\frac{4}{5}$ × 12)

E) 1:5   12 eggs = 2 broken ($\frac{1}{6}$ × 12); 10 unbroken ($\frac{5}{6}$ × 12)

24. Elk Herd (🌶)

Choice B. Since the ratio of male elk to female elk is 16 to 9, it is true that 16 of every 25 (i.e, 16 + 9) elk are male. In other words, a part-to-part ratio of 16 to 9 becomes a part-to-whole ratio of 16 to 25. A ratio of 16:25 equals 64 percent (i.e., $\frac{16}{25}$ = 64%).

25. Rum & Coke (🌶🌶)

Choice B. The trap answer is choice A because it erroneously adds component parts of the two different ratios. That is:

$1:2 + 1:3$ does not equal $2:5$. This could only be correct if ratios represent identical volumes. We cannot simply add two ratios together unless we know the numbers behind the ratios.

	Total	Rum	Coke
Solution #1	6	2	4
Solution #2	32	8	24
Totals		10	28
Final Ratio		5	14

Simplify ☺

The ratio of $10:28$ simplifies to $5:14$.

Supporting Calculations:

$6 \times \frac{1}{3} = 2$ Two ounces of rum in solution #1.

$6 \times \frac{2}{3} = 4$ Four ounces of Coke in solution #1.

$32 \times \frac{1}{4} = 8$ Eight ounces of rum in solution #2.

$32 \times \frac{3}{4} = 28$ Twenty-eight ounces of Coke in solution #2.

26. Toledo Prep (🌶🌶)

Choice C. Likely the best way to approach this problem is to pick numbers. A faculty-to-student ratio of 1 to 7 means that for every 80 persons, 10 are faculty members and 70 are students:

Breakdown: $80 \times \frac{1}{8} = 10$ faculty members; $80 \times \frac{7}{8} = 70$ students

Keeping with these same numbers, if 10 people are faculty members, then 2 are female and 8 are male: ($10 \times \frac{1}{5} = 2$) females and $(10 \times \frac{4}{5} = 8)$ males. For every 70 students, 30 are female and 40 are male:

Breakdown: $70 \times \frac{3}{7} = 30$ females; $70 \times \frac{4}{7} = 40$ males

Therefore: $\dfrac{\text{female}}{\text{total persons}} = \dfrac{2 + 30}{10 + 70} = \dfrac{32}{80} = \dfrac{4}{10} = 40\%$

Another way to organize the information in this problem is with a matrix. Again, assume that the total number of students and faculty is 80.

	Faculty	Students	
Male	8	40	48
Female	2	30	32
	10	70	80

27. Fuchsia (🌱🌱🌱)

Choice D. This is a more complex problem than Toledo Prep, but conceptually similar. We know that there are 24 liters of Fuchsia in a ratio of 5 parts red to 3 parts blue. The first step then is to break the 24 liters of fuchsia into actual amounts of "red" and "blue" paint. This requires using a part-to-whole ratio (i.e., three-eighths blue and five-eighths red; $\frac{3}{8}$ and $\frac{5}{8}$ respectively).

Blue: 5 parts red to 3 parts blue:

$$\frac{3}{5+3} = \frac{3}{8} \quad \rightarrow \quad \frac{3}{8} \times 24 = 9 \, \text{liters (of blue paint)}$$

Red: 5 parts red to 3 parts blue:

$$\frac{5}{5+3} = \frac{5}{8} \quad \rightarrow \quad \frac{5}{8} \times 24 = 15 \text{ liters (of red paint)}$$

Our final ratio is a part-to-part ratio, comparing red and blue paint in fuchsia to the red and blue paint in mauve. So the final formula, expressed as a proportion, becomes:

$$\frac{15_{\text{red}}}{9_{\text{blue}} + x_{\text{blue}}} = \frac{3_{\text{red}}}{5_{\text{blue}}}$$

$5(15) = 3(9 + x)$
$75 = 3(9 + x)$
$75 = 27 + 3x$
$48 = 3x$
$3x = 48$
$x = 16$ liters (of blue paint)

28. Exchange ( )

Choice D. Since one Philippine peso equals 0.15 Hong Kong dollars, one Hong Kong dollar divided by 0.15 equals 6.67 pesos.

Ratio Structure:

$$\frac{A}{B} :: \frac{C}{D}$$

$$\frac{\text{Part}}{\text{Whole}} :: \frac{\text{Part}}{\text{Whole}}$$

Applied to this problem:

$$\frac{\text{Philippine peso}}{\text{Hong Kong dollar}} :: \frac{\text{Philippine peso}}{\text{Hong Kong dollar}}$$

Calculation:

$$\frac{1_{PP}}{0.15_{HK\$}} = \frac{x_{PP}}{1_{HK\$}}$$

$1(1) = 0.15(x)$
$1 = 0.15x$
$0.15x = 1$
$$x = \frac{1}{0.15}$$
$x = 6.67$ pesos

Author's note: In the above equation, we simply reverse the equation and "$1 = 0.15(x)$" becomes "$0.15x = 1$." The rationale behind this practice may be viewed in two ways. The first is based in logic. If A equals B, then it is true that B equals A. The second is based in algebra and unfolds as follows:

$1 = 0.15x$
$-0.15x = -1$ [switching sides changes the signs]
$(-1)(-0.15x) = (-1)(-1)$ [multiply both sides by -1]
$0.15x = 1$

In short, when solving for a single variable such as x, our practical goal is to get x on one side of the equation and all other terms on the opposite side of the equation. Typically, this involves isolating x on the left-hand side of the equation while placing all the other terms on the right-hand side of the equation. There are two useful concepts to keep in mind when manipulating elements of a formula to in order to achieve this objective. The first is that bringing any number (or term) across the equals sign, changes the sign of that number (or term). That is, positive numbers become negative and negative numbers become positive. The second useful concept is that whatever we do to one term in an equation, we must do to every other term in that same equation (in order to maintain the equivalent value of all terms in the equation). So in the above equation we may choose to multiply each side of the equation by −1 in order to cancel the negative signs. We do this because we always want to solve for a positive x value.

29. Sale Time (ℐ)

Choice D. The quick method is to divide $12 by 30%. That is, $12 ÷ 0.30 = $40. The ratio and proportion method follows below:

$$\frac{\$12}{30\%} = \frac{x}{100\%}$$

$$100\%(\$12) = 30\%(x)$$

$$30\%(x) = 100\%(\$12)$$

$$\frac{1}{30\%} \times 30\%(x) = \frac{1}{30\%} \times 100\%(\$12)$$

$$x = \frac{\$12(100\%)}{30\%}$$

$$x = \frac{\$12(1.0)}{0.3}$$

$$x = \$40$$

Note, per the penultimate step above, we may choose to convert percentages (i.e., 100% and 30%) to decimals (i.e., 1.0 and 0.3) for ease of computation.

Extra: Suppose a problem asked: "A man bought a pair of shoes for $28 during a 30% off sale. What was the retail price of the shoes?"

Again, the quick method is to divide $28 by 70%. That is, $28 ÷ 0.70 = $40. This is the same problem but viewed from a different angle. Here's the more formal ratio-and-proportion method:

$$\frac{\$28}{70\%} = \frac{x}{100\%}$$

$$100\%(\$28) = 70\%(x)$$

$$70\%(x) = 100\%(\$28)$$

$$\frac{1}{\cancel{70\%}} \times \cancel{70\%}\,(x) = \frac{1}{70\%} \times 100\%(\$28)$$

$$x = \frac{\$28(100\%)}{70\%}$$

$$x = \frac{\$28(1.0)}{0.7}$$

$$x = \$40$$

30. Shrinkage (💲💲)

Choice C. Arguably the fastest way to solve this problem is to set it up as a ratio or proportion.

$$\frac{75\%}{100\%} = \frac{24}{x}$$

$$75\%(x) = 100\%(\$24)$$

$$\frac{1}{\cancel{75\%}} \times \cancel{75\%}\,(x) = \frac{1}{75\%} \times 100\%(\$24)$$

$$x = \frac{\$24(100\%)}{75\%}$$

$$x = \frac{\$24(1.0)}{0.75}$$

$$x = 32 \text{ ounces}$$

31. Product A ( )

Choice D.

$$\frac{120\%}{100\%} = \frac{x}{\$40}$$

$$120\%(\$40) = 100\%(x)$$

$$100\%(x) = 120\%(\$40)$$

$$\frac{1}{\cancel{100\%}} \times \cancel{100\%}\,(x) = \frac{1}{100\%} \times 120\%(\$40)$$

$$x = \frac{\$40(120\%)}{100\%}$$

$$x = \frac{\$40(1.2)}{1.0}$$

$$x = \$48$$

Note that we cannot divide $40 by 0.8 to get $50. This is incorrect because the reciprocal of 1.2 is 0.83, not 0.8. As illustrated by tip #11, $\frac{120\%}{100\%}$ is not the same as $\frac{100\%}{80\%}$.

32. Product B ( )

Choice C.

$$\frac{120\%}{100\%} = \frac{\$50}{x}$$

$$120\%(x) = 100\%(\$50)$$

$$\frac{1}{\cancel{120\%}} \times \cancel{120\%}(x) = \frac{1}{120\%} \times 100\%(\$50)$$

$$x = \frac{\$50(100\%)}{120\%}$$

$$x = \frac{\$50(1.0)}{1.2}$$

$$x = \$41.67$$

33. Product C ( )

Choice E.

$$\frac{100\%}{80\%} = \frac{x}{\$40}$$

$$100\%(\$40) = 80\%(x)$$

$$80\%(x) = 100\%(\$40)$$

$$\frac{1}{\cancel{80\%}} \times \cancel{80\%}(x) = \frac{1}{80\%} \times 100\%(\$40)$$

$$x = \frac{\$40(100\%)}{80\%}$$

$$x = \frac{\$40(1.0)}{0.8}$$

$$x = \$50$$

34. Product D (🌶)

Choice B.

$$\frac{100\%}{80\%} = \frac{\$50}{x}$$

$$100\%(x) = 80\%(\$50)$$

$$\frac{1}{100\%} \times 100\%(x) = \frac{1}{100\%} \times 80\%(\$50)$$

$$x = \frac{\$50(80\%)}{100\%}$$

$$x = \frac{\$50(0.8)}{1.0}$$

$$x = \$40$$

35. Diners (🌶 🌶)

Choice B. This nice, round number represents the cost before tax and tip.

Calculation of the cost <u>before</u> tip:

$$\frac{x}{\$264} = \frac{100\%}{120\%}$$

$$120\%(x) = 100\%(\$264)$$

$$\frac{1}{120\%} \times 120\%(x) = \frac{1}{120\%} \times 100\%(\$264)$$

$$x = \frac{\$264(100\%)}{120\%}$$

$$x = \frac{\$264(1.0)}{1.2}$$

$x = \$220$

Calculation of the cost <u>before</u> tip <u>and</u> taxes:

$$\frac{x}{\$220} = \frac{100\%}{110\%}$$

$$110\%(x) = 100\%(\$220)$$

$$\frac{1}{\cancel{110\%}} \times \cancel{110\%}(x) = \frac{1}{110\%} \times 100\%(\$220)$$

$$x = \frac{\$220(100\%)}{110\%}$$

$$x = \frac{\$220(1.0)}{1.1}$$

$$x = \$200$$

Author's note: The quick method is to divide $264 by 1.2 and then by 1.1. That is, $[(\$264 \div 1.2) \div 1.1)] = \$200$. Likewise, we can divide $264 by 1.32. That is, $\$264 \div (1.2 \times 1.1) = \$264 \div 1.32 = \$200$.

36. Investments (🜪🜪)

Choice E. Below are the calculations for gain and loss expressed as mathematical proportions.

Gain on Sale of Property A:

$$\frac{120\%}{100\%} = \frac{\$24,000}{x}$$

$$120\%(x) = 100\%(\$24,000)$$

$$\frac{1}{\cancel{120\%}} \times \cancel{120\%}(x) = \frac{1}{120\%} \times 100\%(\$24,000)$$

$$x = \frac{\$24,000(100\%)}{120\%}$$

$$x = \frac{\$24,000(1.0)}{1.2}$$

$$x = \$20,000$$

Gain: $24,000 – $20,000 = $4,000

This gain represents the sales price less original purchase price.

Loss on the Sale of Property B:

$$\frac{100\%}{80\%} = \frac{x}{\$24,000}$$

$$100\%(\$24,000) = 80\%(x)$$

$$80\%(x) = 100\%(\$24,000)$$

$$x = \frac{\$24,000(100\%)}{80\%}$$

$$x = \frac{\$24,000(1.0)}{0.8}$$

$$x = \$30,000$$

Loss: $30,000 – $24,000 = $6,000

This loss represents the original purchase price less the amount received from the sale.

Therefore, we have an overall loss of $2,000 (net $6,000 loss and $4,000 gain). Note that the following provides shortcut calculations:

Calculation of gain:

$24,000 ÷ 1.2 = $20,000
Gain: $24,000 – $20,000 = $4,000

Calculation of loss:

$24,000 ÷ 0.8 = $30,000
Loss: $30,000 − $24,000 = $6,000

Again, we have an overall loss of $2,000 (net $6,000 loss and $4,000 gain).

37. Deceptive (🌶)

Choice D. Dividing 100 by 0.75 is the same as multiplying 100 by the reciprocal of 0.75. The reciprocal of 0.75 is 1.33, not 1.25! This problem highlights the importance of tip #12 (see chapter 1).

38. Topsy-Turvy (🌶)

Choice C. Multiplying a number by 1.2 is the same as dividing that same number by the reciprocal of 1.2, that is, 0.83. Note how answer choice B offers a tempting but incorrect answer of 0.80.

39. Partners (🌶)

Choice E. Incorrect answer choices A, B, C, and D all highlight the commutative property of multiplication—they are equivalent expressions for $a(bcd)$. Correct answer choice E is not an equivalent expression of $a(bcd)$; this answer choice spotlights an incorrect use of the distributive property. The distributive property may be summarized as follows: $ab + ac = a(b + c)$ and $ab − ac = a(b − c)$. Examples: $2 \times 8 + 2 \times 3 = 2(8 + 3)$ and $2 \times 8 − 2 \times 3 = 2(8 − 3)$. The distributive property applies to multiplication over addition and subtraction; it does not apply to multiplication over multiplication or division over division.

40. Bargain ( )

Choice C. This example highlights the commutative property of multiplication. It doesn't matter in which order we multiply numbers, for the answer remains the same. In this case, a 10% discount followed by a 30% discount is the same as selling a given product for 90% of its original price multiplied by 70 percent of its original price (90% × 70% = 63%). A 30 percent discount followed by a 10% discount is also the same as selling a given product for 70% of its original price times 90% of its original price (70% × 90% = 63%).

Contrast this problem with *Double Discount* (problem 9). *Bargain* does not require the actual amount of overall discount, but rather the relationship between the two discounts. For the record, a 10% discount followed by a 30% discount is the same as one 37% discount (not 40% as offered by trap answer choice B).

CHAPTER 2 – WONDERFUL MATH RECIPES

41. Country Club ( )

Choice A.

Two-Groups Formula		Solve
Add:	Group A (Tennis)	700
Add:	Group B (Golf)	500
Less:	Both	<?>
Add:	Neither	1,500
	Total	2,500

Calculation:

Group A + Group B – Both + Neither = Total
700 + 500 – x + 1,500 = 2,500

$$2,700 - x = 2,500$$
$$x = 200$$

That is, 200 people play <u>both</u> tennis and golf. Below is the visual or Venn-diagram representation for this problem.

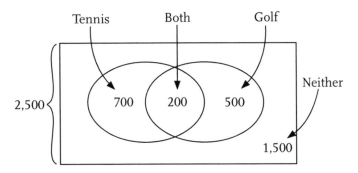

42. Nova vs. Rebound (🦶🦶)

Choice D. We cannot assume that a person liked either Nova or Rebound without taking into account the possibility that a person could like both name brand shoes at the same time. This situation results in overlap and accounts for the fact that the percentage appears to be greater than 100% (i.e., 53% + 47% + 24% = 124%).

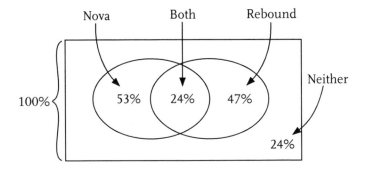

Although not required, the following is the mathematical solution to this problem.

Two-Groups Formula		Solve
Add:	Group A (Nova)	53%
Add:	Group B (Rebound)	47%
Less:	Both	$<x>$
Add:	Neither	24%
	Total	100%

Calculation:

Group A + Group B – Both + Neither = Total
53% + 47% – x + 24% = 100%
124% – x = 100%
x – 24%

Therefore, the number of people liking both Nova and Rebound is 24%. This problem is both strange and tricky because 24% turns out to be the identical figure for the "both" and "neither" components in the group formula. Although this is indeed a peculiar result, it is certainly a mathematical possibility.

43. Science ( )

Choice B. Six people are enrolled in both chemistry and physics.

Two-Groups Formula		Solve
Add:	Group A (Chemistry)	22
Add:	Group B (Physics)	19
Less:	Both	$<x>$
Add:	Neither	0
	Total	35

Calculation:

Group A + Group B − Both + Neither = Total
$22 + 19 - x + 0 = 35$
$41 - x = 35$
$x = 6$

Below is an alternative algebraic calculation. This alternative algebraic approach works because everyone is enrolled in either chemistry or physics or both, and there is no possibility of someone not being enrolled in either. Here the variable b represents those students enrolled in both chemistry and physics.

$(22 - b) + b + (19 - b) = 35$
$41 - b = 35$
$b = 6$

44. Standardized Test (🌶️ 🌶️)

Choice A.

Two-Groups Formula		Solve
Add:	Group A (1st Question)	85%
Add:	Group B (2nd Question)	75%
Less:	Both	<65%>
Add:	Neither	x%
	Total	100%

Calculation:

Group A + Group B − Both + Neither = Total
$85\% + 75\% - 65\% + x = 100\%$
$95\% + x = 100\%$
$x = 5\%$

Five percent of the test takers got neither the first question nor the second question correct. In the Venn-diagram pictorial below, note that 100 percent is what is in the "box"; it includes Q1 & Q2 and "neither," but it must not include "both" because "both" represents overlap that must be subtracted out; otherwise the overlap will be double counted. In other words, either Q1 may include the 65% as part of its 85% or Q2 may include the 65% as part of its 75%, but Q1 and Q2 cannot both claim it. Another way to view this problem is to break up analytically as follows: The number of students who only got Q1 correct is 20% (85% − 65%) while the number of students who only got Q2 correct is 10% (75% − 65%). Since 65% of students got both questions correct, the number of students who got one or the other question correct is: 20% + 10% + 65% = 95%; 5% of students got neither question correct.

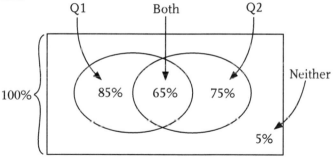

45. Language Classes (🌶🌶🌶)

Choice D. This problem is more difficult and merits a three-chili rating insofar as it requires an answer that is expressed in terms of an algebraic expression.

Use the classic "two-groups" formula:

Group A + Group B − Both + Neither = Total

Applied to the problem at hand:

Spanish + French − Both + Neither = Total Students

$$S + F - B + N = X$$
$$N = X + B - S - F$$

Therefore, expressed as a percent:

$$\text{Neither} = 100 \times \frac{X + B - S - F}{X}$$

46. Valley High (🌶🌶🌶)

Choice B.

Three-Groups Formula	Solve
Add: Biology	4
Add: Chemistry	4
Add: Physics	4
Less: Biology & Chemistry	<2>
Less: Biology & Physics	<2>
Less: Chemistry & Physics	<2>
Add: Biology & Chemistry & Physics	1
Add: None of Biology, Chemistry or Physics	0
Total	x

Calculation:

$$B + C + P - BC - BP - CP + BCP + None = Total$$
$$4 + 4 + 4 - 2 - 2 - 2 + 1 + 0 = x$$
$$x = 7$$

Seven students are involved in independent study projects. The diagram below confirms that there are seven students, as indicated by letters standing for names of students. According to our calculation above, student G (who is taking all of Biology, Chemistry, and Physics) needs to be added back because he or she has been subtracted out one too many times as a result of deducting the three double overlaps.

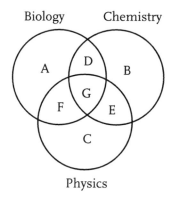

Biology   Chemistry

Physics

47. German Cars (🌶🌶🌶)

Choice C. There are two ways to solve this problem. One employs the "three-groups" formula, while the other involves using the Venn-diagram approach. The three-groups formula is preferable because it is clearly fastest, relying directly on the numbers found right in the problem. In short, the total number of individuals owning a single car are added together, the number of individuals owning exactly two of these cars is subtracted from this number, and, finally, the number of people owning all three of these cars are added back. The calculation follows:

Three-Groups Formula		Solve
Add:	BMW	45
Add:	Mercedes	38
Add:	Porche	27
Less:	BMW & Mercedes	<15>
Less:	BMW & Porche	<12>
Less:	Mercedes & Porche	<8>
Add:	BMW & Mercedes & Porche	5
Add:	None of BMW & Mercedes & Porche	0
	Total	80

Calculation:

B + M + P − BM − BP − MP + BMP + None = Total
45 + 38 + 27 − 15 − 12 − 8 + 5 + 0 = x
x = 80

The Venn-diagram approach is highly analytical and requires breaking the problem down into non-overlapping areas while finding individual values for all seven areas. That's right!—seven distinct areas are created when three groups overlap.

27 + 16 + 12 + 10 + 7 + 3 + 5 = 80

BMW only + Mercedes only + Porsche only − [(BMW & Mercedes) − (Mercedes & Porsche) − (BMW & Porsche)] + (BMW & Mercedes & Porsche).

48. Job Search (🌶)

Choice C. Eleven of the candidates have at least seven years' work experience and hold degrees.

Step #1: Sketch a matrix and enter given information into the appropriate boxes. The question mark depicts the value we're trying to find.

	< 7 years' work experience	≥ 7 years' work experience	
With degrees		?	23
No degrees	3		
		20	35

Step #2: Let's total the numbers on the side and bottom of the matrix.

	< 7 years' work experience	≥ 7 years' work experience	
With degrees		?	23
No degrees	3		12
	15	20	35

Step #3:   Since data must total down and across, we simply fill in remaining numbers within the middle four boxes.

	< 7 years' work experience	≥ 7 years' work experience	
With degrees	12	11	23
No degrees	3	9	12
	15	20	35

49. Nordic ( )

Choice A. Five percent of the people in this Norwegian town have neither blond hair nor blue eyes. Note that since we are dealing with percentages, the total in the box in the bottom right-hand corner is 100%.

Step #1:   Sketch a matrix and enter given information into the appropriate boxes.

	Blond Hair	Non Blond Hair	
Blue Eyes	45%		65%
Non Blue Eyes		?	
	75%		100%

Step #2: Total the numbers on the side and bottom of the matrix.

	Blond Hair	Non Blond Hair	
Blue Eyes	45%		65%
Non Blue Eyes		?	35%
	75%	25%	100%

Step #3: Since data must total down and across, we simply fill in remaining numbers within the middle four boxes.

	Blond Hair	Non Blond Hair	
Blue Eyes	45%	20%	65%
Non Blue Eyes	30%	5%	35%
	75%	25%	100%

50. Singles (🮂🮂)

Choice D. Two-thirds of the women are single (i.e., $\frac{20}{30} = \frac{2}{3}$). For this problem, assume for simplicity's sake that there are 100 students in the course and fill in the given information, turning percentages into numbers.

51. Batteries (🌶🌶🌶)

Choice A. To obtain the percentage of defective batteries sold by the factory, we fill in the information per the matrix below to obtain $\frac{3}{75}$ or $\frac{1}{25}$ or 4%. As in the previous problem, the technique of picking the number "100" greatly simplifies the task at hand.

First, plug the given data from the problem into the matrix:

	Defective	Not Defective	
Rejected		$\frac{1}{10}(80) = 8$	$\frac{1}{4}(100) = 25$
Not Rejected	?		?
	$\frac{1}{5}(100) = 20$	80	100

Second, fill in and complete the matrix:

	Defective	Not Defective	
Rejected	17	8	25
Not Rejected	3	72	75
	20	80	100

52. Experiment ()

Choice A. Let's pick numbers. Say the total number of rats is 100, of which 60 are female and 40 are male. Let's say that 50 rats die, which means 15 were female and 35 were male. Calculation for dead rats: 30% × 50 = 15 female rats versus 70% × 50 = 35 male rats.

There are two ways to solve this problem: the "picking numbers" approach and the "algebraic" approach.

I.   Picking Numbers Approach

Let's pick numbers. Say the total number of rats that are originally alive is 100, of which 60 are female and 40 are male. Let's say that 50 rats die, which means 15 were female and 35 were male. Calculation for dead rats is: 30% × 50 = 15 female rats versus 70% × 50 = 35 male rats.

	Male	Female	
Rats that Died	35	15	*50
Rats that Lived			
	40	60	100

Note that "*50" is merely a guesstimate. However, upon completing the matrix, we are able to calculate the ratio of the death rate among male rats to the death rate among female rats:

	Male	Female	
Rats that Died	35	15	*50
Rats that Lived	5	45	**50
	40	60	100

Note that "**50" is also an estimate and a plug number. Data about the number of rats that lived is not useful in the problem at hand; the focus is on the number of rats that died. Using numbers in the completed matrix, the final calculation follows:

$$\frac{\dfrac{\text{male rat deaths}}{\text{male rats (total)}}}{\dfrac{\text{female rat deaths}}{\text{female rats (total)}}} = \frac{\dfrac{35}{40}}{\dfrac{15}{60}} = \frac{35}{40} \times \frac{60}{15} = 7:2$$

II. Algebraic Approach

$$\frac{\dfrac{0.7 \text{ died (male)}}{0.4 \text{ total (male)}}}{\dfrac{0.3 \text{ died (female)}}{0.6 \text{ total (female)}}} = \frac{0.7 \text{ died}}{0.4 \text{ total}} \times \frac{0.6 \text{ total}}{0.3 \text{ died}} = \frac{0.42^{7}}{0.12^{2}} = \frac{7}{2} = 7:2$$

The trap answer, per choice B, may be calculated in two ways. The first involves using numbers we previously obtained through the picking numbers approach. In this case: $\frac{35}{15} = \frac{7}{3} = 7:3$. However, this ratio is not correct because it involves dividing the estimated number of dead male rats by the number of dead female rats. We need to instead divide the death rate of male rats by the death rate of female rats.

Another way of obtaining trap answer choice B is as follows:

$$\frac{\dfrac{\text{male rat deaths}}{\text{male rats (total)}}}{\dfrac{\text{female rat deaths}}{\text{female rats (total)}}} = \frac{\dfrac{70\% \times 40\%}{40\%}}{\dfrac{30\% \times 60\%}{60\%}} = \frac{\dfrac{28\%}{40\%}}{\dfrac{18\%}{60\%}} = \frac{28\%}{40\%} \times \frac{60\%}{18\%} = 7:3$$

The problem here involves multiplying the respective death rates for male and female rats by the percentage of male and female rats. However, this is erroneous since we do not know how many rats died. In other words, we are dealing with two different groups of rats. The first group represents the total number of rats and the second group represents the rats that died. No direct link can be drawn between these two groups so we cannot simply multiply these percentages.

53. Petroleum (🎈🎈)

Choice C. This is a wet mixture. We need to find the <u>percentage</u> of <u>oil</u> in a final solution.

Next we create an equation from the totals (in bold) appearing at the bottom of the barrels:

$$40\%(50) + 20\%(25) = 75x$$
$$20 + 5 = 75x$$
$$25 = 75x$$
$$75x = 25$$

$$x = \frac{25}{75}$$

$$x = 33\frac{1}{3}\%$$

54. Perfect Blend (🌶️🌶️)

Choice B. This is a dry mixture. We need to calculate the <u>amount</u> of <u>coffee</u> in a second mixture that needs to be added to a first mixture to arrive at a final mixture.

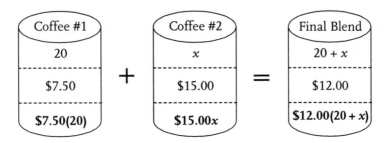

$7.50(20) + \$15.00(x) = \$12.00(20 + x)$

$150 + \$15.00(x) = \$240 + \$12.00x$

$3.00x = \$90$

$$x = \frac{\$90}{\$3.00} \quad \text{(Note: dollar signs cancel)}$$

$x = 30$ pounds

55. Nuts (🌶️🌶️)

Choice A. This is a dry mixture. We need to calculate the <u>amounts</u> of <u>two</u> different <u>nut</u> mixtures to arrive at a final mixture.

Peanuts		Cashews		Final Mix
x		$100 - x$		100
$\$1.50$	$+$	$\$4.00$	$=$	$\$2.50$
$\$1.50(x)$		$\$4.00(100-x)$		$\$2.50(100)$

$\$1.50(x) + \$4.00(100 - x) = \$2.50(100)$

$\$1.50x + \$400 - \$4.00x = \$250$

$-\$2.50x + \$400 = \$250$

$-\$2.50x = -\$150$

$(-1)(-\$2.50x) = (-1)(-\$150)$ [multiply both sides by -1]

$\$2.50x = \$150$

$x = \dfrac{\$150}{\$2.50}$

$x = 60$

Therefore, there are 60 pounds of peanuts and 40 pounds of cashews (i.e., $100 - 60 = 40$ pounds of cashews).

56. Adding H$_2$O (𝄡 𝄡)

Choice B. This is a wet mixture. We need to calculate the amount of pure water that needs to be added to arrive at a final solution. Note that the percentage of water added is 0%. Pure water lacks any mixture. This causes the middle term in the equation to drop out.

Water/Salt #1		Water/Salt #2		Final Mix
40		x		$40 + x$
30%	$+$	0%	$=$	25%
$30\%(40)$		$0\%(x)$		$25\%(40 + x)$

$$30\%(40) + 0\%(x) = 25\%(40 + x)$$
$$12 + 0 = 10 + 0.25x$$
$$2 = 0.25x$$
$$0.25x = 2$$
$$x = \frac{2}{0.25}$$

$x = 8$ gallons of water

57. Evaporation (🌶🌶🌶)

Choice E. This is a wet mixture. We need to calculate the amount of <u>pure water</u> that needs to be <u>subtracted</u> to arrive at a final solution. The percentage of the water subtracted is 0 percent. Pure water lacks any mixture. Again, this is the reason the middle term in the equation drops out.

Water/Sugar		Water		Final Mix
50		x		$50 - x$
3%	−	0%	=	10%
3%(50)		0%(x)		10%(50 − x)

$$3\%(50) - 0\%(x) = 10\%(50 - x)$$
$$1.5 - 0 = 5 - 0.10x$$
$$1.5 - 5 = -0.10x$$
$$-3.5 = -0.10x$$
$$(-1)(-3.5) = (-1)(-0.10x) \quad \text{[multiply both sides by } -1]$$
$$3.5 = 0.10x$$
$$0.10x = 3.5$$
$$x = \frac{3.5}{0.10}$$

$x = 35$ liters of water

58. Gold ( )

Choice E. This is a dry mixture. We need to calculate the <u>amount</u> of <u>pure gold</u> that needs to be added to arrive at a final alloy.

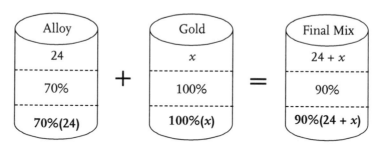

$$70\%(24) + 100\%(x) = 90\%(24 - x)$$
$$16.8 + x = 21.6 + 0.90x$$
$$1.0x - 0.90x = 21.6 - 16.8$$
$$0.10x = 4.8$$
$$x = \frac{4.8}{0.1}$$
$$x = 48 \text{ ounces of pure gold}$$

Author's note: When we add pure gold, we use 100%. Thus the term at the bottom of the middle barrel becomes $100\%(x)$. If we were to add a pure non-gold alloy, then the percentage would be 0 and the middle term would drop out in our calculation.

59. Bowling ( )

Choice B. Because the number of games bowled is equal, we can simply perform weighted average to arrive at the answer.

Weighted Average = (Event$_1$ × Weight$_1$) + (Event$_2$ × Weight$_2$)

$$WA = 110(60\%) + 130(40\%)$$
$$WA = 66 + 52$$
$$WA = 118$$

The alternative format below is user-friendly when performing calculations.

$$\text{Event}_1 \times \text{Weight}_1 = xx \qquad\qquad 110 \times 60\% = 66$$
$$\text{Event}_2 \times \text{Weight}_2 = \underline{xx} \qquad\qquad 130 \times 40\% = \underline{52}$$
$$\underline{\underline{xx}} \qquad\qquad\qquad \underline{\underline{118}}$$

60. The Right Mix (🌶)

Choice C.

Weighted Average = (Event$_1$ × Weight$_1$) + (Event$_2$ × Weight$_2$)

WA = \$900(80%) + \$15.00(20%)
WA = \$7.20 + \$3.00
WA = \$10.20

This is de facto a weighted average problem. As such, we simply multiply dollars (events) by their respective percentages (weights).

61. Baseball (🌶🌶)

Answer A.

Weighted Average = (Event$_1$ × Weight$_1$) + (Event$_2$ × Weight$_2$)

WA = (3% × 0.5) + (5% × 0.5)
WA = 1.5% + 2.5%
WA = 4%

If the price of one baseball bat increases by 3%, then the price of three baseball bats also increases by 3% (not 9% or 12%). The percentage change in aggregate price (cost) is the same regardless of the number of baseball bats in question. Ditto for a catcher's mitt.

Since a baseball bat and a catcher's mitt currently cost the same, the baseball bat and catcher's mitt each cost one-half of the total price. So one half of the total price has increased by 3%, while the other half has increased by 5%. Therefore, the total price has increased by 4%.

62. Venture Capital (🌶)

Choice B. Expected return or expected earnings is calculated using the weighted average formula for three events.

The weighted average formula can be summarized as follows:

$$WA = (E_1 \times W_1) + (E_2 \times W_2) + (E_3 \times W_3)$$

$$WA - \$5,000,000(20\%) + \$1,500,000(50\%) + \$500,000(30\%)$$
$$WA = \$1,000,000 + \$750,000 + \$150,000$$
$$WA - \$1,900,000$$

The following format is "computation-friendly":

$$\$5,000,000 \times 20\% = \$1,000,000$$
$$\$1,500,000 \times 50\% = \$750,000$$
$$130 \times 40\% = \underline{\$150,000}$$
$$\underline{\underline{\$1,500,000}}$$

63. Portfolio (🌶 🌶)

Choice A. Weighted average, again, is the basis for calculating the expected return.

$$\text{Expected Return} = WA_1 \times WA_2 + WA_3$$

First Investment:

$$WA_1 = (E_1 \times W_1) + (E_2 \times W_2) + (E_3 \times W_3)$$

$$WA_1 = \$90,000\left(\tfrac{1}{6}\right) + \$50,000\left(\tfrac{1}{2}\right) + - \$60,000\left(\tfrac{1}{3}\right)$$

$$WA_1 = 15,000 + 25,000 + -20,000$$

$$WA_1 = 20,000$$

Second Investment:

$$WA_2 = (E_1 \times W_1) + (E_2 \times W_2)$$

$$WA_2 = \$100,000\left(\tfrac{1}{2}\right) + - \$50,000\left(\tfrac{1}{2}\right)$$

$$WA_2 = \$50,000 + -25,000$$

$$WA_2 = 25,000$$

Third Investment:

$$WA_3 = (E_1 \times W_1) + (E_2 \times W_2) + (E_3 \times W_3) + (E_4 \times W_4)$$

$$WA_3 = \$100,000\left(\tfrac{1}{4}\right) + \$60,000\left(\tfrac{1}{4}\right) + (-\$40,000)\left(\tfrac{1}{4}\right) + (-\$80,000)\left(\tfrac{1}{4}\right)$$

$$WA_3 = \$25,000 + \$15,000 + -\$10,000 + -\$20,000$$

$$WA_3 = 10,000$$

So, the expected return on all three investments is:

Expected Return = $20,000 + $25,000 + $10,000
Expected Return = $55,000

Chapter 3 – Favorite Numeracy Dishes

64. Gross Margin (🌶)

Choice B.

$$\text{Margin \%} = \frac{\text{Sales (\$)} - \text{Cost (\$)}}{\text{Sales (\$)}} \times 100\%$$

$$\text{Margin \%} = \frac{\$10 - \$7.50}{\$10} \times 100\%$$

$$\text{Margin \%} = \frac{\$2.50}{\$10} \times 100\%$$

$$\text{Margin \%} = 0.25 \times 100\% = 25\%$$

65. Markup (🌶)

Choice C.

$$\text{Markup \%} = \frac{\text{Sales (\$)} - \text{Cost (\$)}}{\text{Cost (\$)}} \times 100\%$$

$$\text{Markup \%} = \frac{\$10 - \$7.50}{\$7.50} \times 100\%$$

$$\text{Markup \%} = \frac{\$2.50}{\$7.50} \times 100\%$$

$$\text{Markup \%} = 0.33 \times 100\% = 33\frac{1}{3}\%$$

66. Alpha Inc. (🌶🌶)

Choice C.

$$\text{Margin \%} = \frac{\text{Markup \%}}{100\% + \text{Markup \%}}$$

$$\text{Margin}\% = \frac{50\%}{100\% + 50\%}$$

$$\text{Margin}\% = \frac{50\%}{150\%}$$

$$\text{Margin}\% = 33\frac{1}{3}\%$$

Author's note: The calculation below is technically more accurate. The percentage signs (i.e., %) cancel out and 0.33 must be multiplied by 100% in order to turn this decimal into a percentage and to reinstate the percentage sign in the final answer.

$$\text{Margin}\% = \frac{50\%}{150\%} = 0.33 \times 100\% = 33\frac{1}{3}\%$$

67. Beta Corp. (🦶 🦶)

Choice B.

$$\text{Markup}\% = \frac{GM\%}{100\% - GM\%}$$

$$\text{Markup}\% = \frac{20\%}{100\% - 20\%}$$

$$\text{Markup}\% = \frac{20\%}{80\%}$$

$$\text{Markup}\% = 25\%$$

Again, the following calculation is more accurate. The percentage signs (i.e., %) cancel out and 0.25 must be multiplied by 100% in order to turn this decimal into a percentage and to reinstate the percentage sign in the final answer.

$$\text{Markup}\% = \frac{20\%}{80\%} = 0.25 \times 100\% = 25\%$$

68. Century Bus Line (🌶)

Choice D. Profit is the interplay of three variables—price, cost, and volume. The key is to review the profit formula to determine which variables are increasing or decreasing in value:

$$(\text{Price}_{\text{per unit}} - \text{Cost}_{\text{per unit}}) \times \text{No. of units} = \text{Profit}$$

Maximum profit will be obtained when price per unit goes up, costs per unit go down, and sales volume goes up. This may be summarized as follows:

$$(\uparrow \text{Price}_{\text{per unit}} - \downarrow \text{Cost}_{\text{per unit}}) \times \uparrow \text{No. of units} = \uparrow\uparrow \text{Profit}$$

In this problem, we know that volume (i.e., the number of paying passengers) is forecasted to increase. Therefore, we must look at the unknown variables, namely price per unit (i.e., ticket sales price per passenger) and total cost per unit (i.e., cost per passenger) as follows:

$$(? \text{Price}_{\text{per unit}} - ? \text{Cost}_{\text{per unit}}) \times \uparrow \text{No. of units} = \downarrow \text{Profit}$$

In summary, either the price per unit of ticket sales will decrease or total costs per unit will increase. No answer choice gives us the option of an increase in costs. Therefore, it must be that price per unit (ticket sales price per passenger) will go down significantly to more than offset the increase in sales volume. A significant reduction in the price of fares would have the effect of reducing overall profits, and 25% would be considered a significant reduction. Answer choices A, B, and C are essentially irrelevant to this scenario as well, and answer choice E would actually serve to increase forecasted profits.

69. Ocean Racer (🌶🌶)

Choice B. The cost is $g + h$ cents per mile. We must multiply this by 100 miles and divide by 100 cents to get dollars. Doing so will result in the original $g + h$.

Total Cost = $(\text{Cost}_1 + \text{Cost}_2) \times$ Number of units

Total Cost = $(g + h) \dfrac{\text{cents}}{\text{~~miles~~}} \times 100 \text{ ~~miles~~}$

Total Cost = $100(g + h)$ cents

Total Cost = $\text{~~100~~}(g + h) \text{ ~~cents~~} \times \dfrac{1 \text{ dollar}}{\text{~~100~~ ~~cents~~}}$

Total Cost = $(g + h)$ dollars

70. Cases (🌶🌶)

Choice B. The profit formula is applicable. Note that the price charged may be viewed as total revenue.

Revenue − Cost = Profit

Revenue = Cost + Profit

Revenue = $\dfrac{1,000(5)}{100} + \dfrac{9,000(y)}{100} + x$

Revenue = $50 + 90y + x$

Author's note: It is easy, when doing these calculations quickly, to forget not only to add x at the end (which represents the expected profit) but also to divide cost components by 100 in order to convert cents to dollars.

71. Delicatessen (🌶️🌶️)

Choice B. A variation of the profit formula is:

$$\text{Profit} = (\text{Price}_{\text{per unit}} \times \text{No. of units}) - (\text{Cost}_{\text{per unit}} \times \text{No. of units})$$

$$\text{Profit} = s\frac{\text{dollars}}{\text{pound}} \times \left(p_{\text{ pounds}} - d_{\text{ pounds}}\right) - \left(c\frac{\text{dollars}}{\text{pound}} \times p_{\text{ pounds}}\right)$$

$$\text{Profit} = s(p - d) - cp$$

Author's note: Gross profit is sales revenue minus product cost. Sales revenue is price per unit multiplied by the number of units: $s(p - d)$. Product cost is cp. Therefore, gross profit is $s(p - d) - cp$. The algebraic expressions cp and pc are, of course, identical.

72. Paper Clips (🌶️)

Choice A. In this problem, cost per unit is c/p. The volume or unit figure is q. Total cost is the product of these two variables.

$$\text{Total Cost} = \text{Cost}_{\text{per unit}} \times \text{Number of units}$$

$$\text{Total Cost} = \frac{c \text{ cents}}{p \text{ clips}} \times q \text{ clips}$$

$$\text{Total Cost} = \frac{cq}{p} \text{ cents}$$

73. Garments (🌶️)

Choice B. This problem requires us to calculate number of units or volume. Start with the basic cost formula and solve for number of units as follows:

$$\text{Total Cost} = \text{Cost}_{\text{per unit}} \times \text{Number of units}$$

$$\text{Therefore, Number units} = \frac{\text{Total Cost}}{\text{Cost}_{\text{per unit}}}$$

$$\text{Number of units} = \frac{{}^{t}\,\text{dollars}}{\frac{d\,\text{dollars}}{s\,\text{shirts}}}$$

$$\text{Number of units} = t\,\cancel{\text{dollars}} \times \frac{{}^{s}\,\text{shirts}}{d\,\cancel{\text{dollars}}}$$

$$\text{Number of units} = \frac{ts}{d}\,\text{shirts}$$

In summary, since the price per unit equals d/s, we divide t by d/s to arrive at ts/d.

74. Copy (🌶)

Choice C.

$$\text{Total Cost} = \text{Cost}_{\text{per unit}} \times \text{Number of units}$$

$$\text{Total Cost} = 2\,\frac{\text{cents}}{\text{page}} \times (y\text{ pages} \times y\text{ pages})$$

$$\text{Total Cost} = 2\,\frac{\text{cents}}{\cancel{\text{page}}} \times y^2\,\cancel{\text{pages}}$$

$$\text{Total Cost} = 2y^2\,\text{cents}$$

Another approach is to first calculate the cost of each manuscript and multiply this expression by the number of manuscripts. Thus, the cost to copy one manuscript ($2y$), is multiplied by the number of manuscripts (y) in order to yield the cost of y manuscripts—$2y^2$.

The answer is simply $2y$; that is, 2 cents per page times y number of pages per manuscript. Then we multiply this number, $2y$, by the number of manuscript copies y to get $2y(y) = 2y^2$.

75. Pete's Pet Shop (🌶🌶)

Choice B. This problem requires that we calculate a usage per unit figure.

$$\text{Cups per bird per day} = \frac{\dfrac{35}{15}\dfrac{\text{cups}}{\text{birds}}}{7\,\text{days}}$$

$$\text{Cups per bird per day} = \frac{35}{15}\frac{\text{cups}}{\text{birds}} \times \frac{1}{7\,\text{days}}$$

$$\text{Cups per bird per day} = \frac{1}{3}\text{ cups per bird per day}$$

After 6 birds are sold, there are 9 birds left. Therefore, the number of cups of bird seed to feed 9 birds for 12 days is:

$$\frac{1}{3}\text{ cup per bird per day} \times 9\text{ birds} \times 12\text{ days} = 36\text{ cups}$$

76. Book Publisher (🌶🌶)

Choice E. This is a classic break-even problem. Note that this problem is not asking for how much profit but rather for the break-even point (i.e., the point of zero profit, achieved when variable revenue is exactly equal to fixed costs or expenses). Moreover, fixed costs and variable costs differ in that variable costs vary directly with sales whereas fixed costs are the same regardless of whether you sell a single item or 1 million items.

Algebraic Method:

Let x be the number of units. The algebraic method is based on a variation of the break-even point formula. Here the break-even point occurs exactly where total costs equal total revenue. Total costs equal fixed costs plus variable costs; total revenue equals sales price multiplied by the number of units sold. By setting the following two formulas equal to one another, we arrive at the break-even point.

Total Revenue = Total Costs

Total Revenue = Fixed Costs + Variable Costs

Total Revenue = FC + VC$_{\text{per unit}}$ × No. of units)

Price$_{\text{per unit}}$ × No. of units = FC + (VC$_{\text{per unit}}$ × No. of units)

Calculation:

$5.95x = (\$150,000 + \$20,00 + \$230,000) + (0.33 + 0.12 + 0.50)x$

$5.95x = \$400,000 + \$0.95x$

$5.95x - \$0.95x = \$400,000$

$5.00x = \$400,000$

$5.00x = \$400,000$

$$x = \frac{\$400,000}{\$5.00}$$

$x = 80,000$ copies (units)

Accountant's Method (per tip #21, chapter 3):

$$\text{BE}_{\text{units}} = \frac{\text{Fixed Costs}}{\text{Selling Price}_{\text{per unit}} - \text{Variable Cost}_{\text{per unit}}}$$

Calculation:

$$\text{Break Even}_{\text{units}} = \frac{\$400,000}{\$5.95 - \$0.95}$$

$$\text{Break Even}_{\text{units}} = \frac{\$400,000}{\$5.00}$$

$$\text{Break Even}_{\text{units}} = 80,000 \text{ copies (units)}$$

Author's note: When dividing $400,000 by $5 per copy, the dollar signs cancel, leaving copies (or units).

77. Sabrina to Change Jobs (🌶🌶)

Choice C. The difference between Sabrina's current base salary, $85,000, and $45,000 is $40,000. Divide $40,000 by 15%($1,500) to get 177.77 unit sales. In the equation below, x stands for the number of sales.

$$\text{Revenue}_{\text{Option 1}} = \text{Revenue}_{\text{Option 2}}$$

$$\$85,000 = \$45,000 + 0.15(\$1,500)(x)$$
$$\$85,000 - \$45,000 = 0.15(\$1,500)(x)$$
$$\$40,000 = 0.15(\$1,500)(x)$$
$$\$40,000 = \$225(x)$$
$$\$225(x) = \$40,000$$
$$x = \frac{\$40,000}{\$225}$$

$$x = 177.77$$
Therefore, $x = 178$ unit sales

Author's note: When dividing $40,000 by $225 (commission) per sale, the dollar signs cancel, leaving number of sales (or units).

Don't be tricked by tempting wrong answer choice B. A total of 177 sales isn't enough to break even. This number must be rounded up to 178 in order to avoid losing money. The actual number of sales is discrete, and can only be represented by whole numbers, not decimals.

With reference to tip #21 (chapter 3), the $40,000 in this problem may be viewed as the fixed costs while the $225(0.15 × $1,500) figure may be viewed as the variable revenue per unit. Thus, $40,000 divided by $225 is equal to 177.77 in unit sales.

78. Grapes ( )

Choice B. The assumption is that both crops were grown on the same number of acres. Of course, per acre costs in each U.S. state are determined as follows:

$$\frac{\text{Costs}}{\text{Acres}} = \text{Cost per acre}$$

In choice B, we cannot make an assessment on the cost per acre basis without knowing the total costs and the total acres. Despite the greater total cost to grow watermelons in Oklahoma (compared with grapes in California), if many more acres of watermelons were grown in Oklahoma (as compared with grapes in California), then the cost per acre to grow watermelons in Oklahoma might very well be lower.

Answer choices A and C fall outside the scope of the argument. In choice A, we are talking about costs to grow, not profits on the sale. Choice C switches terms around. In the original, we are only talking about watermelons in Oklahoma and grapes in California. We must stick with the information as presented. In choice D, how the crops are being used is irrelevant. In choice E, the source of money used to grow crops is also irrelevant. It does not matter who pays for the cost of growing the crops—farmers, the government (state subsidies), or Aunt Jessie—costs are costs.

79. Act-Fast ( )

Choice D. This problem is built on the assumption that the number of tablets per bottle is equal. Therefore, if a bottle of aspirin contains more than twice as many tablets as does a bottle of Act-Fast, then logic dictates that a bottle of aspirin provides a greater amount of pain reliever. And since the cost per bottle is the same, it stands to reason that the cost efficiency of aspirin is greater than that of Act-Fast.

What we really want to determine is how much pain reliever we're getting and how much it is costing us. In choice A, there is no mention of the number of tablets per bottle, so we cannot determine how much pain reliever in aggregate we're getting for our money. Furthermore, implied logic dictates that if one Act-Fast tablet contains twice the pain reliever found in a tablet of regular aspirin then an Act-Fast tablet should be twice the size of a regular aspirin tablet (or an aspirin tablet should be one-half the size of an Act-Fast tablet). Whereas choice B is beyond the scope of the argument, choice C weakens the argument but not nearly to the extent choice D does. There is no need for the pain reliever to be of a different type or strength, as suggested by choice E.

80. Prototype ( )

Choice B. This type of problem highlights the concepts of "efficiency" and "cost efficiency."

First, we set up the problem conceptually:

$$\left(1.0 \, \frac{\text{dollar}}{\text{liter}} \times 1.0 \, \text{liter}\right) - \$x = \left(1.2 \, \frac{\text{dollar}}{\text{liter}} \times \frac{5}{9} \, \text{liters}\right)$$

Remember that $\$x$ is equal to the dollar savings. Question: Where does the fraction $\frac{5}{9}$ come from? An 80 percent increase in efficiency can be expressed as $\frac{180\%}{100\%}$ or $\frac{9}{5}$. The reciprocal of $\frac{9}{5}$

is $\frac{5}{9}$. This means that the P-Car needs only five-ninths as much fuel to drive the same distance as does the T-Car.

Second, we solve for x which represents the cost savings when using the P-Car.

Let's convert $1.2 to $\frac{6}{5}$ for simplicity.

$$\$(1.0)(1.0) - \$x = \$\left(\frac{6}{5}\right)\left(\frac{5}{9}\right)$$

$$\$1.0 - \$x = \$\left(\frac{6}{5}\right)\left(\frac{5}{9}\right)$$

$$\$1 - \$x = \$\frac{2}{3}$$

$$-\$x = -\$1 + \$\frac{2}{3}$$

$$-\$x = -\$\frac{1}{3}$$

$$(-1)(-\$x) = (-1)(-\$\tfrac{1}{3}) \quad \text{[multiply both sides by } -1]$$

$$\$x = \$\frac{1}{3} \quad \text{[dollar signs cancel]}$$

$$x = \frac{1}{3} \text{ or } 33\frac{1}{3}\%$$

Since we save one-third of a dollar for every dollar spent, our percentage savings is $33\frac{1}{3}$ percent. Therefore, although the cost of gas for the T-Car is more expensive, it results in an overall cost efficiency.

For the record, whereas 80 percent represents how much more <u>efficient</u> the P-Car is compared to the T-Car, the correct answer, $33\frac{1}{3}$ percent represents how much more <u>cost efficient</u> the P-Car is compared to the T-Car.

81. Wanda ( )

Choice C. If we assume that the problems on the test are distributed proportionally or linearly across the four sections of the test, then yes, Wanda has finished more than 75% of the test. She has finished 76 problems out of 100 and is now working on her 77th problem. If the problems on the test are not distributed proportionally, we cannot tell whether she has finished more or less than 75% of the exam.

82. Noble Book Club ( )

Choice B. The argument claims that the free CD-ROM gift program has succeeded in increasing Noble Book Club sales. However, this claim would be greatly weakened if the company sold the exact same number of books but just had more concentrated sales patterns at specific points in time. In other words, instead of 100 people buying one book each, perhaps 25 people bought four books each. The free CD-ROM gift program may have succeeded in selling more books to selected customers, but overall, there might be little change in terms of the total number of books sold.

The sales representative's assumption is that the free CD-ROM gift program is the driving force behind increased sales. Choice A would fall outside the scope of the argument. We are concerned in this argument with the linkage between the CD-ROM gift and sales of four or more books, not with sales of less than four books. Choice C is irrelevant since we are only concerned about the increase in sales, not costs, and not profits. Choice D is irrelevant because what other book clubs did or did not do is of little concern. Choice E only says membership is up in general, but that is not the same thing as saying Noble Book Club sales increased.

83. Max Motorcycles (🌶🌶)

Choice D. If the number of motorcycles built by Max Motorcycles has not increased sharply in the past 10 years, then this will strengthen the argument. But if the number of motorcycles built by Max has increased sharply in the past 10 years, then this will weaken the argument. For it could be that a large number of Max Motorcycles that are on the road today were just recently manufactured and there hasn't been sufficient time to test their durability. Therefore, by knowing that this is not the case, the argument is strengthened. This concept is referred to as back-loading as opposed to front-loading. The argument is strengthened knowing that back-loading (a recent increase in production of motorcycles) is <u>not</u> occurring. Stated another way, the argument is weakened knowing that back-loading is occurring, i.e., a large proportion of new motorcycles has been recently released in the marketplace.

CHAPTER 4 – SPECIAL MATH GARNISHMENTS

84. Matching Exercise (🌶🌶)

Rising Graphs (select from graphs A, B, C or D):

For every additional hour you work, you are paid incrementally more dollars for each hour worked.

B

You get a fixed base salary but may also get additional pay based on hours worked which are not covered by your base salary.

D

For every additional hour you work, you get paid incrementally fewer dollars for each hour worked.

C

For every hour you work, you get paid y dollars per hour.

A

Flat Graphs (select from graphs E, F, G or H):

You got a promotion at work which increased the fixed salary you now receive.

F

Because of layoffs at work, you decided to take a reduction in your fixed monthly salary in order to keep your current job.

G

You get paid a fixed (flat) wage regardless of the hours you work.

E

You get paid by the hour, but cannot earn more than a certain dollar amount during the course of your working week.

H

Falling Graphs (select from graphs I, J, K or L):

Your yearly pension decreases in terms of real dollars during each year of your retirement.

I

Your wages decrease gradually with age until the time you stop working altogether.

K

Your wages are fixed but stop the moment you retire.

L

Your wages decrease dramatically with age but you can still earn a little in your retirement.

J

85. Grand Hotel Q1 (🌶🌶)

Answer: 5%. Blackjack is a type of card and table game. Blackjack accounted for $200,000 worth of total card and table game revenue. Total revenue from all card and table games—blackjack ($200,000), roulette ($500,000), craps ($100,000), and baccarat ($200,000)—amounted to $1,000,000 (see bar chart). Thus blackjack accounted for 20% of card and table game revenue. Since the percentage of total hotel and casino revenue attributable to card and table games was 25% (see pie chart), the final calculation is the product of these two numbers.

$$\frac{\$200,000}{\$1,000,000} = 20\%$$

$$20\% \times 25\% = 5\%$$

Author's note: With respect to bar charts and pie charts used in combination, concentrate on how one graph fits into the other. Because pie charts are best at breaking down information and bar charts are best at ranking information, the total dollar amount of the bar chart will most likely relate to a given percentage of the pie chart.

86. Grand Hotel Q2 (🌶🌶)

Answer: $800,000. Use a ratio and proportion to figure out how much money was attributed to in-house shows. We know that card and table games revenue totals $1 million and represents 25% of the pie. We also know that in-house shows account for 20% of all hotel and casino revenue. Therefore, this relationship provides the key to solving for the amount of money brought in by in-house shows.

$$\frac{25\%}{\$1,000,000} :: \frac{20\%}{x}$$

$$25\%(x) = 20\%(\$1,000,000)$$

$$25\%(x) = \$200,000$$

$$x = \frac{\$200,000}{25\%}$$

$$x = 800,000$$

87. Grand Hotel Q3 (🌶)

Answer: 15%. Again, a ratio and proportion does the trick. The percentage of penthouse suites rental revenues equals the percentage of miscellaneous revenues, so both revenue sources equal 3%.

$$\frac{\% \text{ Penthouse Suites Rentals}}{\% \text{ Total Room Rental \& Restaurant Revenue}} = \frac{3\%}{20\%} = 15\%$$

88. Workforce Q1 (🌶)

Answer: 1.5 million. First, the number of people working in the blue-collar sector in 1970 was 4.5 million; (45% × 10,000,000 = 4.5 million). Second, the number of people working in the blue-collar sector in 2009 was 3 million; (20% × 15,000,000 = 3 million). Therefore, the difference between 4.5 million and 3 million accounts for the 1.5 million decrease in the number of workers in this sector.

Author's note: When viewing two pie charts and comparing the percentage growth of items over two different time frames, beware of directly comparing the percentage figures of one pie chart with the other. To compare two figures, first translate these figures to actual dollars, based on the dollar size of each pie chart, and then make dollar or percentage comparisons.

89. Workforce Q2 (🌶 🌶)

Answer: The Blue-Collar Sector. The number of people working in the service sector in 2009 was 4.5 million (30% of 15,000,000). The number of people working in the blue-collar sector in 1970 was also 4.5 million (45% of 10,000,000). Although the absolute percentage of service sector workers in 2009 is smaller than the absolute percentage of blue-collar workers in 1970, the size of the workforce has grown by 5 million. The pie has gotten bigger. In this case, a smaller percentage of a bigger pie is equal to a larger percentage of a smaller pie.

90. Workforce Q3 (𝕊𝕊𝕊)

Answer: Service, Professional, and Clerical Sectors. All three sectors experienced a higher percentage increase compared with the percentage decrease attributed to the blue-collar sector!

The tricky point here is that we cannot compare percentages directly. In other words, we cannot simply subtract the percentages for any particular sector to arrive at a percentage increase or decrease. For example, the absolute percentage decrease in the number of workers in the blue-collar sector (i.e., 45% – 20% = 25%) is obviously larger than any absolute percentage increase for any other sector. However, this type of calculation is not possible because we're asked for the percentage increase or decrease in the number of workers in each sector. These are relative numbers and we must take into account the fact that the size of the pies we're comparing are not the same. The workforce is larger in 2009.

Service Sector (percentage increase):

$$\frac{New - Old}{Old} = \frac{4.5M - 1.5M}{1.5M}$$

$$\frac{3M}{1.5M} = 2 \times 100\% = 200\%$$

Professional Sector (percentage increase):

$$\frac{New - Old}{Old} = \frac{3.75M - 1.5M}{1.5M}$$

$$\frac{2.25\,M}{1.5\,M} = 1.5 \times 100\% = 150\%$$

Clerical Sector (percentage increase):

$$\frac{New - Old}{Old} = \frac{3.75\,M - 2.5\,M}{2.5\,M}$$

$$\frac{1.25\,M}{2.5\,M} = 0.5 \times 100\% = 50\%$$

Blue-Collar Sector (percentage decrease):

$$\frac{Old - New}{Old} = \frac{4.5\,M - 3\,M}{4.5\,M}$$

$$\frac{1.5\,M}{4.5\,M} = \frac{1}{3} \times 100\% = 33\frac{1}{3}\%$$

91. Magna Fund Q1 (🌶)

Choice C. Refer to the chart *Value of Portfolio*. The y-axis is labeled in increments of $50 million. We can estimate the value of the portfolio during Q1 as being $75 million. The value of the portfolio during Q3 sits at $125 million. The difference between these two segments is $50 million. Therefore, $50 million divided by $75 million is equal to a $66\frac{2}{3}\%$ increase.

92. Magna Fund Q2 (🌶 🌶 🌶)

Choice C. Statement I is incorrect. The percentage of cash increases from the third to fourth quarter (Q3 to Q4). Although we see a black downward-sloping line, don't be deceived. This means that the percentage of cash is getting larger. In Q4, cash amounts to fully 50% of the value of the portfolio. When reading a cumulative line graph, concentrate on the space or area covered by each item in the cumulative line graph rather than on whether the lines enclosing these given areas are pointing upward or downward.

Statement II is also incorrect. Bonds accounted for one-third or more of the portfolio in exactly two of four quarters, namely the III and IV quarters. Note that one-third is $33\frac{1}{3}$%, not 30%; bonds only accounted for exactly 30% of the portfolio in Q1, not percent.

Statement III is correct. The average amount of dollar investment per individual investor was greatest during the fourth quarter. Don't be deceived by the fact that Q3 has the greatest portfolio value and the greatest number of investors. Average means total dollars divided by total investors. So lots of money divided by fewer investors means rising investment per individual investor. Throughout the first, second, and third quarters the investment per individual investor remains relatively stable. In the fourth quarter the value of the portfolio falls and so does the number of investors. However, the number of investors falls proportionately greater, meaning that more money is divided among fewer investors. This is very much an exercise in fractions. A denominator that increases incrementally faster than the numerator of a fraction results in a larger fraction. Actual or approximate numbers are as follows:

Quarter 1:

$$\frac{\$75,000,000}{85,000} = \$882 \text{ per investor}$$

Quarter 2:

$$\frac{\$100,000,000}{100,000} = \$1,000 \text{ per investor}$$

Quarter 3:

$$\frac{\$125,000,000}{110,000} = \$1,136 \text{ per investor}$$

Quarter 4:

$$\frac{\$75,000,000}{40,000} = \$1,875 \text{ per investor}$$

Statement IV is also correct. In the second quarter, stocks account for 60% of the total portfolio (80% less 20%). This means that preferred shares are 30% of 60%, or preferred shares accounted for 18% of the total portfolio. Looking at the *Value of Portfolio* chart, we can see that total portfolio value during Q2 is $100,000,000.

The final calculation follows: 18% x $100,000,000 = $18,000,000. Alternatively, you may prefer working from largest to smallest: ($100,000,000 × 60%) × 30% = $18,000,000.

93. Correlation Exercise ( )

Graph A – negative-high correlation
Graph B – positive-low correlation
Graph C – negative-perfect correlation
Graph D – zero correlation
Graph E – positive-high correlation
Graph F – negative-low correlation
Graph G – positive-perfect correlation
Graph H – non-linear correlation

94. Trends Q1 ( )

Answers: Graphs A and B. The graphs of which individual products show an identical mean, median, and mode?

Both graphs of products A and B have identical means, medians, and modes, all of which are equal to 50 years of age. Note the dotted line from the top of each graph to the x-axis which represents the variable age. A graph which is completely symmetrical will have an identical mean, median, and mode.

95. Trends Q2 ( )

Answer: Graph C. The graph of which of the above four products shows the smallest mode?

Product C has the smallest mode while product D has the largest mode. The mode is the place where each graph peaks because this is the spot where a value appears most frequently. For the record, product C has a mode of 35, product A has a mode of 50, product B has a mode of 50, and product D has a mode of 65.

96. Trends Q3 ( )

Answer: Graph C. The graph of which product shows a greater mean compared with its respective mode?

Visually, the graph of product C is skewed toward the larger values, and the mean is more affected by these larger values than is the median or mode. In statistical parlance, graphs skewed to the right (with a shape such as this graph) will always have a mean that is greater than their median and a median which is, in turn, greater than their mode (i.e., mean > median > mode).

Graph skewed to the right

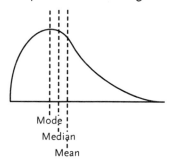

Mode
Median
Mean

A classic real-life situation of a graph skewed to the right (toward larger values) is that of population and income earnings. Most people earn about the same, but a very small percentage of the population earns a tremendous amount, and this discrepancy pulls the mean toward these larger values. The median is also pulled toward these larger values, but less dramatically so.

Author's note: Graphs that are skewed to the right have their tails on the right-hand side. They are so called because they taper off to the right. Visually, the highest point on the graph is situated to the left, and this may make the label "skewed to the right" seem counterintuitive. The exact opposite holds true for graphs that are skewed to the left, as seen in the next pictorial.

97. Trends Q4 (𝄞 𝄞)

Answer: Graph D. The graph of which product shows a greater mode compared with its respective mean?

Visually, the graph of product D is skewed toward left (toward smaller values), and the mean is more affected by the inclusion of these smaller values than is the median or mode. In statistical parlance, graphs skewed to the left (with a shape such as this graph) will always have a mode that is greater than their median, which is in turn greater than their mean (i.e., mean < median < mode).

Graph skewed to the left

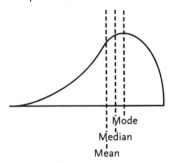

Graphs skewed to the left are less common in real life than are graphs which are skewed to the right. An example of a graph that is skewed to the left would be the age printed on coins. If we were to go looking for coins, most of those we'd find in circulation would be newly minted. If we looked hard, eventually we'd find some very old coins. The number of old coins would create a graph skewed to the left (i.e., number of coins found would be placed on the y axis while the year of coinage, from oldest to newest, would mark the x axis). The mean and median are pulled toward these smaller values, but the median less dramatically so.

98. Trends Q5 (🐍)

Answer: Graph B. The graph of which product shows the smallest range?

Range is defined as the smallest value subtracted from the largest value:

> Product A has the largest range, 80 − 20 = 60 years.
> Product B has a range of 70 − 30 = 40 years.
> Product C has a range of 80 − 20 = 60 years.
> Product D also has a range of 80 − 20 = 60 years.

99. Trends Q6 ()

Answer: Graph B. The graph of which product shows the lowest standard deviation?

Product B has the lowest standard deviation because the values are most bunched around its arithmetic mean. A low standard deviation tells us that the values are bunched around the arithmetic mean; a high standard deviation tells us that the values are spread out. Product A likely has the greatest standard deviation because the values are most spread out on both sides. Graphs C and D depict data that is fairly bunched around the mean.

100. Statistician ( )

Choice D. Statement I is false. The difference between these two numbers (or values) may or may not be deemed important. Laypersons understandably but mistakenly associate the word "significance" with the word "importance." Statisticians use the word "significant" to mean "probable," so the words "statistically significant" translate as "statistically probable."

Statement II is true. The difference between two numbers (or values) may be determined to be statistically significant or not statistically significant. These two determinations are mutually exclusive. It's one or the other, and never both at the same time.

Statement III is true. The statistical odds that the difference between these two numbers or values cannot be relied upon are less than a 5% chance. A 5% chance is the same as a 1 in 20 chance. The words "cannot be relied upon" mean the same as "cannot be believed to be true."

Statement IV is true. There is at least a 95% chance that the difference between these two numbers or values can be relied upon and less than a 5% chance that this difference is due to chance. Voila!—We have cited the very definition of statistical significance. Of course, it expresses a definition both in terms of what is true as well as what is not true. That is, if we have a 95% or greater chance of being right, we also have a less than 5% chance of being wrong. Finally, "relied upon" may be interpreted as "probably true or probably real." So statement IV could have stated: "There is at least a 95% chance that the difference between these two numbers or values is probably true, and a less than 5% chance that this difference is not probably true (or not probably real)."

Quiz – Answers

1. False. The percentage of people who are female at this conference is $\frac{1}{3}$ or $33\frac{1}{3}\%$, not 50%.

   See tip #13, chapter 1.

2. False. For a given product, markup is always larger than margin.

   See tip #18, chapter 3.

3. False. The cost of the meal before tax and tip was $100. Calculation: ($132 ÷ 1.2) ÷ 1.1 = $100.

   See tip #14, chapter 2.

4. False. Ratios tell us nothing about actual size or value; they tell us instead about relative size or value.

   See tip #11, chapter 2.

5. False. Multiplying a number by 1.2 is the same as dividing the number by the reciprocal of 1.2, which is 0.83, not 0.8. Case in point: $100 × 1.2 = $100 ÷ 0.83.

   See tip #16, chapter 2.

6. False. Break-even occurs exactly where variable revenue (sales revenue less variable costs) equals total fixed costs.

   See tip #21, chapter 3.

7. False. The store item is now selling for a 44% discount, or 56% of its original price. For example, $100 less 20% equals $80, and $80 less 30% equals $56. A $44 discount on $100 is 44%. The trap here is that you can't add (or subtract) the percentages of different wholes.

See tip #6, chapter 1.

8. False. Fifteen people bought both hamburgers and hotdogs, as calculated using the formulas below.

    Hamburgers + Hotdogs − Both + Neither = Total
    $60 + 35 - x + 20 = 100$
    $x = 15$

    See tip #1, chapter 1.

9. False. If product A is selling for 20% more than product B, then the ratio of product A to product B is 120% to 100%.

    See tip #15, chapter 2.

10. False. Data with a high standard deviation is spread out. Data with a low standard deviation is bunched.

    See tip #29, chapter 4.

Selected Bibliography

Dewdney, A.K. *200% of Nothing: An Eye-Opening Tour through the Twists and Turns of Math Abuse and Innumeracy.* New York: Wiley, 1993.

Graduate Management Admission Council. *The Official Guide for GMAT Quantitative Review.* 2nd ed. Hoboken, NJ: Wiley, 2009.

Green, Sharon Weiner and Ira K. Wolf. *Barron's GRE.* 17th ed. Hauppauge, NY: Barron's, 2007.

Julius, Edward H. *Rapid Math Tricks and Tips: 30 Days to Number Power.* New York: Wiley, 1992.

Lerner, Marcia. *Math Smart: Essential Math for These Numeric Times.* 2nd ed. New York: Princeton Review Publishing, 2001

"Math." *Math.com: The World of Math Online.* http://www.math.com/.

Nardi, William A. *How to Solve Algebra Word Problems.* 4th ed. New York: Arco, 2002.

OnlineMathLearning. http://www.onlinemathlearning.com/.

Paulos, John Allen. *Innumeracy: Mathematical Illiteracy and Its Consequences.* NewYork: Hill & Wang, 1988.

Stanton, Robert. *Math Power: Score Higher on the SAT, GRE, and Other Standardized Tests.* 3rd ed. NewYork: Kaplan Publishing, 2003.

Stapel, Elizabeth. *Purplemath.* http://www.purplemath.com/.

Stewart, Ian. *Professor Stewart's Cabinet of Mathematical Curiosities.* New York: Basic Books, 2009.

Wikipedia, The Free Encyclopedia, s.v. "Mathematics," http://en.wikipedia.org/wiki/Mathematics (accessed December 5, 2009).

A mind is not a vessel to be filled, but a fire to be kindled.

—Plutarch

Index

Numbers in *italics* (within brackets) indicate multiple-choice problems, 1 to 100. They are preceded by corresponding page numbers.

About the Author

Brandon Royal is an award-winning author who has worked as a marketing analyst, auditor, teacher, corporate trainer, and publisher. As a corporate trainer, he has conducted seminars for such well-known organizations as AT&T, A.T. Kearney, Merrill Lynch, and J.P. Morgan. In a former life, he worked for the Ernst & Young Accounting Group in Hong Kong, conducting joint-venture audits for client companies including Coca-Cola and the Hilton and Holiday Inn hotels in China. Currently he is pursuing his chosen occupation—authorship—and oversees the publication and distribution of his own line of books.

Brandon developed an expertise in test preparation and honed his theories of teaching and education while working in Hong Kong for US-based Kaplan Educational Centers—a Washington Post subsidiary and the oldest and largest test-preparation organization in the world. A graduate of the University of Chicago's Graduate School of Business, his interest in writing began after completing fiction and scriptwriting courses at Harvard University. Since then he has authored several books and written across a diverse spectrum of genres, penning fiction, nonfiction, short stories, essays, news articles, speeches, poetry, and personal quotations. Reviews of his books have appeared in Time Asia magazine, Publishers Weekly, Library Journal of America, Midwest Book Review, The Asian Review of Books, Choice Reviews Online, Asia Times Online, and About.com.

To contact the author:
E-mail: contact@brandonroyal.com
Web site: www.brandonroyal.com

Books by Brandon Royal

The Little Red Writing Book:
20 Powerful Principles of Structure, Style and Readability

The Little Gold Grammar Book:
Mastering the Rules That Unlock the Power of Writing

The Little Green Math Book:
30 Powerful Principles for Building Math and Numeracy Skills

The Little Blue Reasoning Book:
50 Powerful Principles for Clear and Effective Thinking

Secrets to Getting into Business School:
100 Proven Admissions Strategies to Get You Accepted at the MBA Program of Your Dreams

Chili Hot GMAT:
200 All-Star Problems to Get You a High Score on Your GMAT Exam

Chili Hot GMAT: Math Review

Chili Hot GMAT: Verbal Review

Bars of Steel:
Life before Love in a Hong Kong Go-Go Bar – The True Story of Maria de la Torre

Pleasure Island:
You've Found Paradise, Now What? A Modern Fable on How to Keep Your Dreams Alive

LaVergne, TN USA
04 February 2011
215317LV00002B/18/P